MOVING APPLICATIONS TO THE CLOUD ON WINDOWS AZURE™

3RD EDITION

Moving Applications to the Cloud on Windows Azure™

3rd Edition

Dominic Betts
Alex Homer
Alejandro Jezierski
Masashi Narumoto
Hanz Zhang

978-1-62114-020-7

Contents

Foreword – Yousef Khalidi

Microsoft's Windows Azure platform, an operating environment for developing, hosting, and managing cloud-based services, established a foundation that allows customers to easily move their applications from on-premises locations to the cloud. With Windows Azure, customers benefit from increased agility, a very scalable platform, and reduced costs.

Microsoft's cloud strategy has three broad tenets: to offer flexibility of choice for deploying services based on business needs; to provide enterprise-level services with no compromises on availability, reliability, or security; and to support consistent, connected experiences across devices and platforms. Windows Azure is a key component of Microsoft's cloud strategy.

Windows Azure builds on Microsoft's many years of experience running online services for millions of users, and on our long history of building platforms for developers. We focused on making the transition from on-premises to the cloud easy for both programmers and IT professionals. Their existing skills and experience are exactly what they need to start using the Windows Azure platform.

Microsoft is committed to Windows Azure, and will continue to expand it as we learn how all our customers around the globe, from the largest enterprises to the smallest ISVs, use it. One of the advantages of an online platform is that it allows us to introduce innovations quickly.

I'm excited to introduce this guide from the Microsoft patterns & practices team, proof of our commitment to help customers be successful with the Windows Azure platform. Whether you're new to Windows Azure, or if you're already using it, you'll find guide a great source of things to consider. I encourage you to get started exploring Microsoft's public cloud and to stay tuned for further guidance from the patterns & practices team.

FOREWORD FOR THE THIRD EDITION

Since its first beginnings, and since I reviewed the original edition of this guide from the patterns & practices team, Windows Azure has continued to mature by offering exciting new services and capabilities. Now that we have achieved general release, with a comprehensive SLA, we have seen a huge uptake of the platform across all sectors of our industry.

In my original foreword I talked about our commitment to the enterprise. We have proved not only that we can deliver on these commitments, but go beyond them to offer even more innovative features; including many that make migration of existing on-premises applications to the cloud much easier. The business case for Windows Azure continues to prove itself, and there is even more to come!

Sincerely,
Yousef Khalidi
Distinguished Engineer, Windows Azure

Preface

How can a company's applications be scalable and have high availability? To achieve this, along with developing the applications, you must also have an infrastructure that can support them. For example, you may need to add servers or increase the capacities of existing ones, have redundant hardware, add logic to the application to handle distributed computing, and add mechanisms to handle failover. You have to do this even if an application is in high demand for only short periods of time. Life becomes even more complicated (and expensive) when you start to consider issues such as network latency and security boundaries.

The cloud offers a solution to this dilemma. The cloud is made up of interconnected servers located in various data centers. However, you see what appears to be a centralized location that someone else hosts and manages. By shifting the responsibility of maintaining an infrastructure to someone else, you're free to concentrate on what matters most: the application. If the cloud has data centers in different geographical areas, you can move your content closer to the people who are using it most. If an application is heavily used in Asia, have an instance running in a data center located there. This kind of flexibility may not be available to you if you have to own all the hardware. Another advantage to the cloud is that it's a "pay as you go" proposition. If you don't need it, you don't have to pay for it. When demand is high, you can scale up, and when demand is low, you can scale back.

Yes, by moving applications to the cloud, you're giving up some control and autonomy, but you're also going to benefit from reduced costs, increased flexibility, and scalable computation and storage. This guide shows you how to do this.

WHO THIS BOOK IS FOR

This book is the first volume in a series about Windows Azure. It demonstrates how you can adapt an existing, on-premises ASP.NET application to one that operates in the cloud. The book is intended for any architect, developer, or information technology (IT) professional who designs, builds, or operates applications and services that are appropriate for the cloud. Although applications do not need to be based on the Microsoft Windows operating system to work in Windows Azure or written using a .NET language, this book is written for people who work with Windows-based systems. You should be familiar with the Microsoft .NET Framework, Microsoft Visual Studio, ASP.NET, and Microsoft Visual C#.

WHY THIS BOOK IS PERTINENT NOW

In general, the cloud has become a viable option for making your applications accessible to a broad set of customers. In particular, Windows Azure now has in place a complete set of tools for developers and IT professionals. Developers can use the tools they already know, such as Visual Studio, to write their applications. In addition, the Windows Azure SDK includes the compute emulator and the storage emulator. Developers can use these to write, test, and debug their applications locally before they deploy them to the cloud. There are also tools and an API to manage your Windows Azure accounts. This book shows you how to use all these tools in the context of a common scenario—how to adapt an existing ASP.NET application and deploy it to Windows Azure.

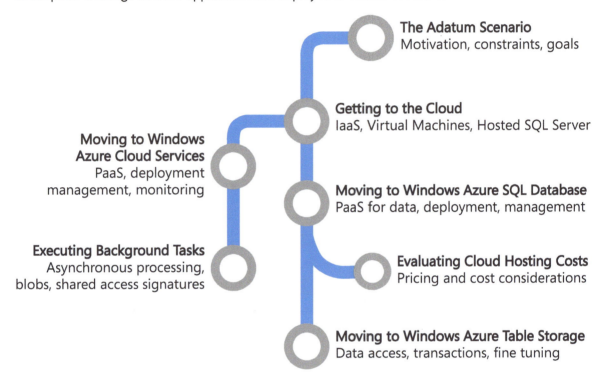

The Adatum Scenario
Motivation, constraints, goals

Getting to the Cloud
IaaS, Virtual Machines, Hosted SQL Server

Moving to Windows Azure Cloud Services
PaaS, deployment management, monitoring

Moving to Windows Azure SQL Database
PaaS for data, deployment, management

Executing Background Tasks
Asynchronous processing, blobs, shared access signatures

Evaluating Cloud Hosting Costs
Pricing and cost considerations

Moving to Windows Azure Table Storage
Data access, transactions, fine tuning

HOW THIS BOOK IS STRUCTURED

Chapter 1, "The Adatum Scenario" introduces you to the Adatum company and the aExpense application. The following chapters describe how Adatum migrates the aExpense application to the cloud. Reading this chapter will help you understand why Adatum wants to migrate some of its business applications to the cloud, and it describes some of its concerns. It will also help you to understand basic options for hosting applications and services in the cloud.

Chapter 2, "Getting to the Cloud" describes the first steps that Adatum took in migrating the aExpense application. Adatum's goal here is simply to get the application working in the cloud, but this includes "big" issues such as security and storage. The chapter shows how Adatum used Windows Azure virtual machines and network services to deploy and communicate with the hosted servers.

Chapter 3, "Moving to Windows Azure Cloud Services" describes how Adatum adapted the aExpense application to run as a hosted service in Windows Azure by using the Cloud Services feature. The chapter describes how Adatum modified the solution, converted it to use claims authentication instead of Active Directory, and took advantage of Windows Azure Caching for the session data.

Chapter 4, "Moving to Windows Azure SQL Database" describes how Adatum evaluated the use of Windows Azure SQL Database instead of a hosted SQL Server by exploring the limitations this might impose and the cost savings that it might provide. The chapter then goes in to show how Adatum converted the aExpense application to use Windows Azure SQL Database.

Chapter 5, "Executing Background Tasks" describes adding a worker role to the aExpense application to process scanned receipt images as a background task. It also shows how aExpense uses Windows Azure blob storage for storing these images, and shared access signatures to provide secure access to them.

Chapter 6, "Evaluating Cloud Hosting Costs" introduces a basic cost model for the aExpense application running on Windows Azure and shows how Adatum calculated the estimated annual running costs for the application.

Chapter 7, "Moving to Windows Azure Table Storage" describes how Adatum switched from using a SQL database to using Windows Azure table storage in the aExpense application. It discusses the differences between the two storage models, and shows how Adatum adapted the data access code to work with Windows Azure table storage. This chapter also discusses how Adatum fine-tuned the application after deployment, and the options it is considering for improving the application in the future.

What You Need to Use the Code

These are the system requirements for running the scenarios:

- Microsoft Windows 7 with Service Pack 1, Microsoft Windows 8, Microsoft Windows Server 2008 R2 with Service Pack 1, or Microsoft Windows Server 2012 (32 bit or 64 bit editions).
- *Microsoft .NET Framework version 4.0.*
- *Microsoft Visual Studio* 2010 Ultimate, Premium, or Professional edition with Service Pack 1 installed, or Visual Studio 2012 Ultimate, Premium, or Professional edition.
- *Windows Azure SDK for .NET* (includes the Windows Azure Tools for Visual Studio). See the Release Notes for information on the specific version required.
- *Microsoft SQL Server 2012, SQL Server Express 2012, SQL Server 2008, or SQL Server Express 2008.* See the Release Notes for information on specific versions depending on your operating system.
- *Windows Identity Foundation.* This is required for claims-based authorization.
- *WatiN 2.0.* Open the Properties dialog and unblock the zip file after you download it and before you extract the contents. Place the contents in the **Lib\Watin** folder of the examples.

Other components and frameworks required by the examples are installed using NuGet when you run the solutions. See the Release Notes included with the examples for instructions on installing and configuring them.

Who's Who

As mentioned earlier, this book uses a set of scenarios that demonstrates how to move applications to the cloud. A panel of experts comments on the development efforts. The panel includes a cloud specialist, a software architect, a software developer, and an IT professional. The scenarios can be considered from each of these points of view. The following table lists the experts for these scenarios.

Bharath is a cloud specialist. He checks that a cloud-based solution will work for a company and provide tangible benefits. He is a cautious person, for good reasons.

"Moving a single application to the cloud is easy. Realizing the benefits that a cloud-based solution can offer is not always so straight-forward".

Jana is a software architect. She plans the overall structure of an application. Her perspective is both practical and strategic. In other words, she considers not only what technical approaches are needed today, but also what direction a company needs to consider for the future.

"It's not easy to balance the needs of the company, the users, the IT organization, the developers, and the technical platforms we rely on.

Markus is a senior software developer. He is analytical, detail-oriented, and methodical. He's focused on the task at hand, which is building a great cloud-based application. He knows that he's the person who's ultimately responsible for the code.

"I don't care what platform you want to use for the application, I'll make it work."

Poe is an IT professional who's an expert in deploying and running in a corporate data center. Poe has a keen interest in practical solutions; after all, he's the one who gets paged at 3:00 AM when there's a problem.

"Migrating to the cloud involves a big change in the way we manage our applications. I want to make sure our cloud apps are as reliable and secure as our on-premise apps."

If you have a particular area of interest, look for notes provided by the specialists whose interests align with yours.

Acknowledgments

On March 4th 2010 I saw an email from our CEO, Steve Ballmer, in my inbox. I don't normally receive much email from him, so I gave it my full attention. The subject line of the email was: "We are all in," and it summarized the commitment of Microsoft to cloud computing. If I needed another confirmation of what I already knew, that Microsoft is serious about the cloud, there it was.

My first contact with what eventually became Windows Azure was about three years ago. I was in the Developer & Platform Evangelism (DPE) team, and my job was to explore the world of software delivered as a service. Some of you might even remember a very early mockup I developed in late 2007, called Northwind Hosting. It demonstrated many of the capabilities that the Windows Azure platform offers today. (Watching an initiative I've been involved with since the early days become a reality makes me very, very happy.)

In February 2009, I left DPE and joined the patterns & practices team. My mission was to lead the "cloud program": a collection of projects that examined the design challenges of building applications for the cloud. When the Windows Azure platform was announced, demand for guidance about it skyrocketed.

As we examined different application development scenarios, it became quite clear that identity management is something you must get right before you can consider anything else. It's especially important if you are a company with a large portfolio of on-premises investments, and you want to move some of those assets to the cloud. This describes many of our customers. Therefore, patterns & practices's first deliverable, and an important milestone in our cloud program, was *A Guide to Claims-Based identity and Access Control*.

The Windows Azure platform is special in many ways. One is the rate of innovation. The various teams that deliver all of the platform's systems proved that they could rapidly ship new functionality. To keep up with them, I felt we had to develop content very quickly. We decided to run our projects in two-months sprints, each one focused on a specific set of considerations.

This guide, now fully updated to cover the new capabilities of Windows Azure, mainly covers a migration scenario: how to move an existing application to the Windows Azure platform. As in the claims guide, we've developed a fictitious case study that explains, step by step, the challenges our customers are likely to encounter.

I want to start by thanking the following subject matter experts and contributors to this guide: Dominic Betts, Scott Densmore, Ryan Dunn, Steve Marx, and Matias Woloski. Dominic has the unusual skill of knowing a subject in great detail and of finding a way to explain it to the rest of us that is precise, complete, and yet simple to understand. Scott brought us a wealth of knowledge about how to build scalable Windows Azure applications, which is what he did before he joined my team. He also brings years of experience about how to build frameworks and tools for developers. I've had the privilege of working with Ryan in previous projects, and I've always benefited from his acuity, insights, and experience. As a Windows Azure evangelist, he's been able to show us what customers with very real requirements need. Steve is a technical strategist for Windows Azure. He's been instrumental in shaping this guide. We rely on him to show us not just what the platform can do today but how it will evolve. This is important because we want to provide guidance today that is aligned with longer-term goals. Last but not least, Matias is a veteran of many projects with me. He's been involved with Windows Azure since the very first day, and his efforts have been invaluable in creating this guide.

As it happens with all our written content, we have sample code for most of the chapters. They demonstrate what we talk about in the guide. Many thanks to the project's development and test teams for providing a good balance of technically sound, focused and simple-to-understand code: Masashi Narumoto (Microsoft Corporation), Scott Densmore (Microsoft Corporation), Federico Boerr (Southworks), Adrián Menegatti (Southworks), Hanz Zhang (Microsoft Corporation), Ravindra Mahendravarman (Infosys Ltd.), Rathi Velusamy (Infosys Ltd.).

Our guides must not only be technically accurate but also entertaining and interesting to read. This is no simple task, and I want to thank Dominic Betts, RoAnn Corbisier (Microsoft Corporation), Alex Homer (Microsoft Corporation), and Tina Burden from the writing and editing team for excelling at this.

The visual design concept used for this guide was originally developed by Roberta Leibovitz and Colin Campbell (Modeled Computation LLC) for A Guide to Claims-Based Identity and Access Control. Based on the excellent responses we received, we decided to reuse it for this book. The book design was created by John Hubbard (eson). The cartoon faces were drawn by the award-winning Seattle-based cartoonist Ellen Forney. The technical illustrations were adapted from my Tablet PC mockups by Chris Burns.

All of our guides are reviewed, commented upon, scrutinized, and criticized by a large number of customers, partners, and colleagues. We also received feedback from the larger community through our CodePlex website. The Windows Azure platform is broad and spans many disciplines. We were very fortunate to have the intellectual power of a very diverse and skillful group of readers available to us.

I also want to thank all of these people who volunteered their time and expertise on our early content and drafts. Among those, we want to highlight the exceptional contributions of Jorge Rowies (Southworks), Marcos Castany (Southworks), Federico Boerr (Southworks), Corey Sanders (Microsoft Corporation), Nir Mashkowski (Microsoft Corporation), Ganesh Srinivasan (Microsoft Corporation), Jane Sinyagina (Microsoft Corporation), Rick Byham (Microsoft Corporation), Boris Scholl (Microsoft Corporation), and Paul Yuknewicz (Microsoft Corporation).

I hope you find this guide useful!

Eugenio Pace
Senior Program Manager – *patterns & practices*
Microsoft Corporation

ACKNOWLEDGEMENTS OF CONTRIBUTORS TO THE THIRD EDITION

Windows Azure is an evolving platform. We originally published the first edition of this guide in 2010, demonstrating a basic set of Windows Azure features. I'm now pleased to release the third edition of this guide, which incorporates the latest and greatest features of Windows Azure such as Virtual Machines, Web Sites, Caching, and more. By taking advantage of these new features, you have a lot more options to choose from when migrating your own applications from on-premises to the cloud.

As our scope increased, we also added new community members and industry experts who have provided significant help throughout the development of this edition. I want to acknowledge the exceptional contributions of the following people: Dominic Betts (Content Master), Alex Homer (Microsoft Corporation), Alejandro Jezierski (Southworks), Mauro Krikorian (Southworks), Jorge Rowies (Southworks), Marcos Castany (Southworks), Hanz Zhang (Microsoft Corporation), Rathi Velusamy, RoAnn Corbisier (Microsoft Corporation), Nelly Delgado (Microsoft Corporation), Eugenio Pace (Microsoft Corporation), Carlos Farre (Microsoft Corporation), Trent Swanson (Full Scale 180 Inc.), Ercenk Keresteci (Full Scale 180 Inc.), Federico Boerr, Corey Sanders (Microsoft Corporation), Nir Mashkowski (Microsoft Corporation), Ganesh Srinivasan (Microsoft Corporation), Jane Sinyagina (Microsoft Corporation), Rick Byham (Microsoft Corporation), Boris Scholl (Microsoft Corporation), and Paul Yuknewicz (Microsoft Corporation). I also want to thank everyone who participated in our CodePlex community site.

Masashi Narumoto
Senior Program Manager – *patterns & practices*
Microsoft Corporation
September 2012

1   The Adatum Scenario

This chapter introduces a fictitious company named Adatum. It describes Adatum's current infrastructure, its software portfolio, and why Adatum wants to move some of its applications to Windows Azure. As with any company considering this process, there are many issues to take into account and challenges to be met, particularly because Adatum has not used the cloud before. At the end of this chapter you will see how Adatum explored and evaluated the major requirements for moving its applications to the cloud, and an overview of the migration steps that Adatum followed. The chapters that follow this one show in detail how Adatum modified its expense tracking and reimbursement system, aExpense, at each stage for deployment to Windows Azure.

THE ADATUM COMPANY

Adatum is a manufacturing company of 15,000 employees that mostly uses Microsoft technologies and tools. It also has some legacy systems built on other platforms, such as AS400 and UNIX. As you would expect, Adatum developers are knowledgeable about various Microsoft products, including .NET Framework, ASP.NET, SQL Server, Windows Server, and Visual Studio. Employees in Adatum's IT department are proficient at tasks such as setting up and maintaining Active Directory and using System Center.

Adatum uses many different applications. Some are externally facing, while others are used exclusively by its employees. The importance of these applications ranges from "peripheral" to "critical," with many lying between the two extremes. A significant portion of Adatum's IT budget is allocated to maintaining applications that are either of mid-level or peripheral importance.

Adatum uses mainly Microsoft products, and its developers are knowledgeable about most Microsoft technologies such as Windows, SQL Server, and the .NET Framework.

1

Adatum wants to change this allocation. Its aim is to spend more money on the services that differentiate it from its competitors and less on those that don't. Adatum's competitive edge results from assets, such as its efficient supply chain and excellent quality controls, and not from how effectively it handles its internal email. For example, Adatum wants efficient email but is looking for more economical ways to provide this so that it can spend most of its budget on the systems that directly affect its customers. Adatum believes that one way to achieve this optimization is to selectively deploy applications to the cloud.

Adatum's Challenges

Adatum faces several challenges. Currently, deploying new on-premises applications takes too long, considering how quickly its business changes and how efficient its competitors are. The timeframe for acquiring, provisioning, and deploying even a simple application can be at least several weeks. No matter the application's complexity, requirements must be analyzed, procurement processes must be initiated, requests for proposals may need to be sent to vendors, networks must be configured, and so on. Adatum must be able to respond to its customers' demands more rapidly than the current procedures allow.

Another issue is that much of Adatum's infrastructure is used inefficiently. The majority of its servers are underutilized, and it's difficult to deploy new applications with the requisite service-level agreements (SLAs) to the existing hardware. Virtual machines are appropriate in some cases, but they are not appropriate in all cases. This inefficiency means that Adatum's capital is committed to an underutilized infrastructure when it could be better used elsewhere in the business.

A final issue is that less critical applications typically get less attention from the IT staff. It is only when the application fails or cannot keep up with demand that anyone takes notice. By this time, the problem is expensive to fix, both in terms of IT time and in inefficient use of the users' time.

Adatum wants to focus on the applications, and not on the infrastructure. Adatum believes that by deploying some of its applications to a public cloud such as Windows Azure it can take advantage of economies of scale, promote standardization of its applications, and have automated processes for managing them. Most importantly, Adatum believes that this will make it more effective at addressing its customers' needs, a more effective competitor in the market, and a better investment for its shareholders.

Adatum's Goals and Concerns

One of Adatum's goals is to improve the experience of all users of its applications. At a minimum, applications in the cloud should perform as well as their on-premises counterparts. The hope, though, is that they will perform better. Many of its applications are used more at some times than at others. For example, employees use the salary tool once every two weeks but rarely at other times. They would benefit if the applications had increased responsiveness during peak periods. This sensitivity to demand is known as *dynamic scalability*.

However, on-premises applications that are associated with specific servers don't provide this flexibility. Adatum can't afford to run as many servers as are needed during peak times because this hardware is dormant the rest of the time. If these applications were located in the cloud, it would be easy to scale them depending on the demand.

Another goal is to expand the ways that users can access Adatum's applications. Currently, applications are only accessible from the intranet. Applications that are located in the public cloud are, by definition, available over the Internet. However, the public cloud also raises questions about authentication. Many of Adatum's applications use Windows authentication so that users aren't required to enter application-specific credentials. Adatum is concerned that its users would need special credentials for each application in the public cloud.

While Adatum intends that the aExpense application will perform at least as well in the cloud as it does running in its own data center, the aim is to take advantage of the inherent scalability and reliability of cloud hosting to achieve better overall performance and availability than the current on-premises deployment.

A third goal is that at least some of Adatum's applications should be *portable*. Portability means that the application can be moved back and forth between a hosted data center and an on-premises data center without any modifications to the application's code or its operations. If both options are available, the risks that Adatum incurs if it does use the cloud are reduced.

In addition to its concerns about security, Adatum has two other issues. First, it would like to avoid a massive retraining program for its IT staff. Second, very few of Adatum's applications are truly isolated from other systems. Most have various dependencies. Adatum has put a great of deal effort into integrating its systems, even if not all of them operate on the same platform. It is unsure how these dependencies affect operations if some systems are moved to the public cloud.

Adatum's Strategy

Adatum is an innovative company and open to new technologies, but it takes carefully considered steps when it implements them. Adatum's plan is to evaluate the viability of moving to the cloud by starting with some of its simpler applications. It hopes to gain some initial experience, and then expand on what it has learned. This strategy can be described as "try, learn, fail fast, and then optimize." Adatum has decided to start with its aExpense application.

THE aEXPENSE APPLICATION

The aExpense application allows Adatum's employees to submit, track, and process business expenses. Everyone in Adatum uses this application to request reimbursements. Although aExpense is not a critical application, it is important. Employees can tolerate occasional hours of downtime, but prolonged unavailability isn't acceptable.

Adatum's policy is that employees must submit their expenses before the end of each month. The majority of employees don't submit their expenses until the last two business days. This causes relatively high demands during a short time period. The infrastructure that supports the aExpense application is scaled for average use across the month instead of for this peak demand. As a result, when the majority of employees try to submit their expenses during the last two business days, the system is slow and the employees complain.

The application is deployed in Adatum's data center and is available to users on the intranet. While traveling, employees access it through a VPN. There have been requests for publishing aExpense directly to the Internet, but it's never happened.

The application stores a great deal of information because most expense receipts must be scanned and then stored for seven years. For this reason, the data stores used by aExpense are frequently backed up.

The application is representative of many other applications in Adatum's portfolio so it's a good test case for using the cloud. Moving the aExpense application to Windows Azure will expose many of the challenges Adatum is likely to encounter as it expands the number of applications that it relocates to the cloud.

The aExpense Architecture

Figure 1 illustrates the aExpense architecture.

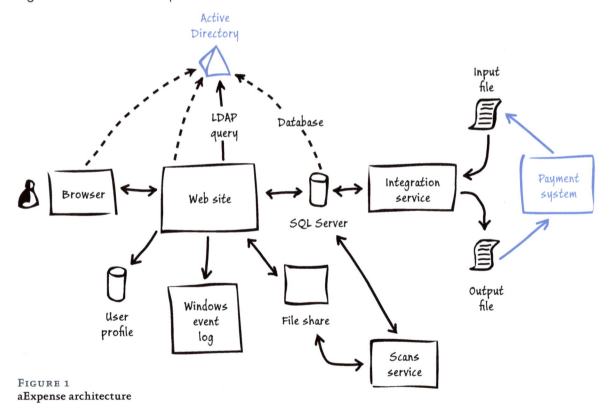

FIGURE 1
aExpense architecture

The architecture is straightforward and one that many other applications use. aExpense is an ASP.NET application and employees use a browser to interact with it. The application uses Windows authentication for security. To store user preferences, it relies on ASP.NET membership and profile providers. Exceptions and logs are implemented with Enterprise Library's Exception Handling Application Block and Logging Application Block. The website uses Directory Services APIs to query for employee data stored in Active Directory, such as the employee's manager. The manager is the person who can approve the expenses.

The aExpense application implements the trusted subsystem to connect to SQL Server. It authenticates with a Windows domain account. The SQL database uses SQL Server authentication mode. The aExpense application stores its information on SQL Server. Scans of receipts are stored on a file share.

There are two background services, both implemented as Windows services. One periodically runs and generates thumbprints of the scanned receipts. It also compresses large images for increased storage efficiency. The other background service periodically queries the database for expenses that need to be reimbursed. It then generates a flat file that the payment system can process. This service also imports the payment results and sends them back to aExpense after the payments are made.

Adatum's aExpense application uses a standard website architecture based on ASP.NET with data stored in SQL Server. However, it does integrate with other in-house systems.

EVALUATING CLOUD HOSTING OPPORTUNITIES

Before initiating a full technical case study for migration of the aExpense application to Windows Azure, the designers and developers at Adatum evaluated the capabilities offered by cloud hosting partner solutions such as Microsoft's Windows Azure. For example, they needed to:

- Identify which type of service offered by the hosting providers best suits Adatum's requirements.
- Determine whether a cloud solution can provide the necessary secure and reliable runtime platform and storage facilities.
- Identify how Adatum can monitor and manage the application
- Determine whether the service level agreements (SLAs) are sufficient to meet Adatum's business requirements.

Evaluating the Runtime Platform

Currently, Adatum runs the aExpense application on its own in-house IT infrastructure. The servers, networks, internal and external connectivity, and associated systems such as power supply and cooling are all the responsibility of Adatum. Together they provide the underlying mechanisms for running applications such as aExpense. As part of the initial evaluation, Adatum investigated the ways that it could move the aExpense application to an external hosting partner.

Infrastructure as a Service

Adatum first considered whether it could simply move the application to an external partner by renting the required infrastructure, complete with all of the associated systems, and run the application unchanged. Renting infrastructure from an external partner is known as *Infrastructure as a Service* (IaaS). Adatum would be responsible for providing and installing the operating system and software, and maintaining it (such as installing operating system and services updates, and upgrading to new versions). The partner company would provide the hardware (the server) and the associated infrastructure and connectivity.

Cloud providers can typically offer very high levels of infrastructure reliability and availability that are beyond the capabilities of many organizations' own datacenters. For example, most incorporate robust disaster recovery processes, and offer the ability to deploy in more than one geographical location.

Adopting an IaaS approach will provide some cost saving through a reduction in overall requirements for in-house infrastructure, but it is not easy (or, in some cases, possible) to quantify the in-house cost of running a specific application. In Adatum's case, the cost of the on-premises infrastructure is effectively shared between all the applications Adatum uses.

In addition, while this approach is attractive, Adatum must take into account the cost of management and maintenance required to keep the hosted operating system running correctly, and the costs of operating system licenses. However, IaaS is generally less expensive than other ways of hosting applications at remote locations. It can also reduce development cost because applications do not need to be refactored to run in specific types of cloud service roles.

Infrastructure now becomes a running cost rather than a capital investment.

IaaS allows you to effectively pick up your server and move it to the cloud with minimal changes required to the application. It is especially useful if you need to deploy on servers that have non-standard configuration, where applications require additional operating system services, or for applications cannot be refactored into a structure suitable for *Platform as a Service* (PaaS) deployment.

Platform as a Service

Secondly, Adatum considered adapting the aExpense application to run as a hosted application on a platform and operating system provided by an external partner. As the application currently runs on Windows Server and uses the .NET Framework, the external partner would need to offer this platform to avoid the costs of porting the application to a different operating system.

Renting a ready-to-use platform from an external partner is known as *Platform as a Service* (PaaS). Adatum would be responsible only for providing and installing its aExpense application, and maintaining it (such as fixing bugs and upgrading to a new version). The partner company would provide the operating system pre-installed on appropriate hardware, with the associated infrastructure and connectivity.

The PaaS approach is attractive to Adatum because it reduces the cost of management and maintenance (the partner is responsible for keeping the operating system running correctly and applying updates), and there is no requirement to pay for operating system licenses. In some cases PaaS hosting charges may be higher than for IaaS, though this is not necessarily the case; and the cost savings in licensing, management, and maintenance can often outweigh any difference. Adatum considered the amount of work involved in refactoring the application to run in cloud-hosted roles and the corresponding development cost, and considered both to be acceptable.

PaaS is particularly useful when applications can be refactored to run using the standard platform offered by cloud hosting providers. Responsibility for managing and updating the operating system and services is delegated to the hosting provider. Applications that use a multi-tier architecture, require administrative access through a virtual network mechanism, or require elevated permissions can be usually be hosted in the cloud using the PaaS model.

Software as a Service

The third option Adatum considered was to abandon their own aExpense application and rent the use of an expenses application provided by another company. Renting use of third party applications is an example of *Software as a Service* (SaaS). Many companies have applications specially designed to handle business expense collation and reporting tasks.

However, Adatum must ensure that the third party application can fully meet its specific requirements; hosted third party applications must typically offer a more generic features set to satisfy a wide range of customers. As well as exploring the overall capabilities of the software, Adatum will need to evaluate its security, configurability, performance, and usability. Changing over may incur costs such as user education, as well as the cost of migrating data and users; and perhaps maintaining the old application for a period until changeover is complete.

Evaluating Data Storage Facilities

Most business applications use data, and so before making any decision about hosting the aExpense application externally Adatum needed to evaluate the data storage and retrieval facilities offered by external partners. On-premises and in-house applications typically use a relational database system based on Structured Query Language (SQL), and Adatum's aExpense application is no exception. Therefore, the external partner must be able to offer the equivalent hosted capability.

However, other storage formats are common. Some applications require storage for disk files or for unstructured data. The aExpense application stores unstructured data in the form of receipt images on a file share, and it also generates disk files for use by other in-house systems. Therefore, the chosen cloud hosting mechanism must be able to provide support for storing unstructured data; this may be in a format other than disk files so long as the application can be easily adapted to use it.

Between them, these mechanisms should be able to provide the data storage and retrieval features that Adatum requires; albeit with some changes to the application code to use the available storage models. By using an appropriate relational database system, or any other type of repository that can be installed on a hosted sever, Adatum can avoid changes to the application code.

> Most business applications rely on a relational database, even though it may be exposed through a custom repository or data access layer. However, many applications also have other storage requirements such as profile and session data, binary and formatted data streams, and disk files. The target hosting platform must either offer equivalent services, or it must be reasonably easy and cost-efficient to adapt the application to use available storage mechanisms.

Evaluating Security, Monitoring, and Management Capabilities

Moving applications to outside of the corporate network prompts several questions not directly related to the hosting platform mechanisms. Adatum must be convinced that the hosting providers' network and infrastructure is secure, and that the hosted application will be protected from malicious attacks and from data exposure in case of systems failure. For example, the hosting network should be resilient to Denial of Service (DoS) and network flooding attacks, and the hosting platform should be able to reliably and safely reinitialize the application after a hardware failure.

In addition, Adatum must evaluate whether hosting in a remote datacenter will meet any legal or regulatory requirements, such as a limitation on the geographical location for data storage and processing. Many cloud hosting providers, including Windows Azure, have datacenters located around the world and allow users to specify the location of the servers and data storage facilities. Windows Azure allows users to specify whether storage replication for backup and resiliency will take place across multiple datacenters in order to satisfy regulatory limitations.

In addition, Adatum must ensure that the chosen hosting provider and deployment mechanism allows administrators to monitor and manage the application and the data stores remotely. Windows Azure includes a range of capabilities that closely match the remote access capabilities for on-premises server, database, and application management. For example, it supports a range of logging facilities, remote desktop access to servers and hosted services, remote configuration, and management of applications and data stores through a web-based portal and APIs that supports REST calls and scripting.

Finally, Adatum must consider if the remote applications must be integrated with other services, both in the cloud and on-premises, to access data, communicate messages, and for monitoring and management. For example, Adatum uses Microsoft System Center Operation Manager for monitoring applications, and it should therefore be also to integrate the remote application and services with this. Additionally, Adatum relies on domain-level authentication through Active Directory and so it will be necessary to join the remote machines to the on-premises domain or adopt an alternative solution that provides equivalent functionality.

Evaluating Service Level Agreements

Adatum recognized that, although the aExpense application is used only by company employees, it must be readily available (in other words, only very rarely offline) and responsive to a reasonable degree. There is no formal SLA for the application, but it should of necessity be available to employees whenever they need to submit expense claims. Of course, for other types of applications, especially publicly visible or business-crucial applications, there will need to be a more formal SLA defined.

SLAs should define not only availability of an application, but also maximum response times. In addition, where other services are required (such as caching or access control), the SLAs should also cover these services. Finally, SLAs should include any information required to define security risks and to meet regulatory or legal requirements (such as the geographical location for data storage).

Windows Azure provides formal SLAs for the IaaS, PaaS, and related services that it offers. However, these do not and cannot cover the customer's hosted application, as this is outside of Microsoft's control. Instead, the SLAs are defined in terms of connectivity and role execution; for example, the SLA for Cloud Services guarantees that a role instance will expose full connectivity for 99.95% of the time and that failed role instances will be detected and restarted 99.9% of the time.

You can find details of the Windows Azure *Service Level Agreements* for all of the services online.

For a full list of the features and services available in Windows Azure, see *"Introducing Windows Azure."*

Evaluating Additional Opportunities

In addition to the fundamental choices of the hosting model and the deployment approach, the designers and developers at Adatum considered if they could benefit from using the many ancillary services and features available in Windows Azure.

For example, they considered whether the application would benefit from the use of Windows Azure Caching to maximize performance when retrieving data; or for caching output, session state, and profile information.

Other features that Adatum realized would be useful for the aExpense application included Windows Azure Active Directory for authentication and the Content Delivery Network (CDN) for delivering images and other non-authenticated content. These features and Adatum's decisions regarding their use are explained in more detail in the following chapters of this guide.

Adatum also considered whether the application needed to communicate with the on-premises applications using messaging, or access services exposed by on-premises applications. Windows Azure Service Bus provides many features that would be useful in this scenario, but Adatum decided that these were not required for the current version of aExpense.

> *To learn more about Windows Azure Service Bus see "Service Bus." The guide "Building Hybrid Applications in the Cloud" describes the scenarios for and usage of Service Bus in detail.*

ADATUM'S MIGRATION PATH FOR THE aEXPENSE APPLICATION

Every company will inevitably make different decisions on the migration path they adopt for moving to the cloud. The range of contributing factors is vast, and each company will have specific goals and limitations that affect the final choices. Typically, companies will begin, as Adatum did, by understanding the concepts of cloud hosting; and then exploring the platforms, services, and options available from cloud hosting providers. From that comes the decision on which cloud provider to use, and the hosting approach that will best match all the requirements.

This guide shows how you can make the appropriate choices when using Windows Azure. However, to help you make those choices, this guide shows several of the hosting approaches. As you will see, the path that Adatum chose for migrating the aExpense application to the cloud included several stages. Adatum began by choosing the option that required the least modification to the aExpense application and then, at each subsequent stage, considered whether moving to another hosting approach would provide additional benefits.

> While the multi-step approach Adatum chose for migrating their application may not be realistic in every real-world scenario, it allows the guide to demonstrate several options that are available for hosting applications in Windows Azure. The discussion of the advantages and limitations at each stage will help you to better understand the options available to you when migrating your own applications.

The migration steps that Adatum took for the aExpense application are shown in the following table. The table shows the chapter that discusses each step, a high-level overview of the options chosen, and the Windows Azure technologies that Adatum used. This will help you to follow the flow of the guide and explore the different approaches taken at each stage.

Chapter	Migration step	Justification	Technologies
2 – "Getting to the Cloud"	Infrastructure as a Service (IaaS).	Minimal code changes to the application and familiarity with the platform. A quick and easy way to explore the benefits of cloud hosting, such as increased reliability and reduced costs of managing the on-premises infrastructure.	Windows Azure Virtual Machines, Virtual Networks, and Connect.
3 – "Moving to Windows Azure Cloud Services"	Platform as a Service (PaaS).	No operating system maintenance, easy scalability and elasticity, more granular control of resource usage, and the opportunity for auto scaling.	Windows Azure Web Sites, Cloud Services web role, and Caching. Windows Identity Framework.
4 – "Moving to Windows Azure SQL Database"	Platform as a Service (PaaS) for database	Lower cost although some limitations on feature availability. No software maintenance.	Windows Azure SQL Database. Transient Fault Handling Application Block.
5 – "Executing Background Tasks"	Maximizing efficiency and adding additional tasks.	Better scalability and performance, better user experience, improved efficiency, and load leveling across role instances.	Windows Azure Cloud Services worker role, queues, and blob storage.
6 – "Evaluating Cloud Hosting Costs"	Revisiting initial cost estimations.	Confirm initial estimates of cost and look for additional savings.	Windows Azure Pricing Calculator.
7 – "Moving to Windows Azure Table Storage"	Switching away from relational database storage.	Lower cost, greater storage volume, opportunity for increased performance, and scalability.	Windows Azure table storage.

Some of the technologies described in this guide and used in the examples are preview versions, and the subsequent release versions may differ from the information provided in this guide. This includes Windows Azure Web Sites, Windows Azure Virtual Machines, and Windows Azure Virtual Networks.

Choosing Your Own Migration Path

Just because Adatum has chosen the path described in this chapter, it doesn't mean that you must follow the same path. Some companies may decide which combination of hosting approach, data store, and services they will use and go directly to this in single migration step. Others may follow a more gradual migration by adopting, for example, Cloud Services as the hosting approach for the application code, but use SQL Server hosted in a Virtual Machine before moving to Windows Azure SQL Database. Meanwhile, some companies may choose the IaaS path so that they have full control over the operating system, but decide to take advantage of the cost savings and vast storage capabilities of Windows Azure table and blob storage instead of using a relational database.

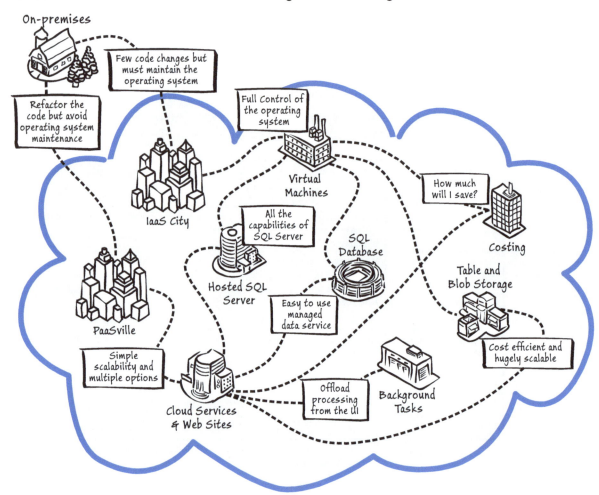

FIGURE 2
Choosing your own migration path

This is one of the major advantages with Windows Azure – you choose which of the wide range of services it offers are most suitable for your own scenario and requirements. No two applications are the same. Throughout this guide you will see more details of the capabilities and limitations of each hosting option, and how to make the right choice for your applications.

MORE INFORMATION

All links in this book are accessible from the book's online bibliography available at: *http://msdn.microsoft.com/en-us/library/ff803373.aspx*.

Overview of Windows Azure features.

For an overview of the data storage options available in Windows Azure, *"Data Storage Offerings on the Windows Azure Platform."*

Introducing Windows Azure includes a list of features.

Windows Azure Service Level Agreements.

"Windows Azure Websites, Cloud Services, and VMs: When to use which?"

Windows Azure Service Bus.

The guide *"Developing Multi-tenant Applications for the Cloud"* explores techniques for building new applications specifically designed for run in Windows Azure.

The guide *"Building Hybrid Applications in the Cloud"* describes the scenarios for and usage of many Windows Azure features.

2 Getting to the Cloud

This chapter describes the first step the developers at Adatum took on their migration path to the cloud. It discusses the contributing factors for the decision they made to use an IaaS approach for hosting the aExpense application in Windows Azure, and explores the process they followed to achieve this. The chapter also discusses issues related to application lifecycle management (ALM), specifically for scenarios that use an IaaS hosting approach.

This first migration step is concerned only with getting the application to work in the cloud without losing any functionality. However, it does address some "big" issues such as security and data storage that are relevant to almost every cloud-based application.

This chapter doesn't explore how to improve the application by exploiting the extended set of features available in Windows Azure. In addition, the on-premises version of the application that you'll see is not complete; for example, it does not support uploading receipt images or interaction with Adatum's other on-premises systems. The following chapters discuss how to improve the application by using other features available in Windows Azure, and you'll see more functionality added to the application. For now, you'll discover how Adatum takes its first steps into the cloud.

THE PREMISE

The existing aExpense application is a business expense submission and reimbursement system used by Adatum employees. The application is built with ASP.NET 4.0, deployed in Adatum's data center, and is accessible from the Adatum intranet. The application relies on Active Directory to authenticate employees. It also uses Active Directory to access some of the user profile data that the application requires; for example, an employee's cost center and manager. Other user profile data is accessed using the ASP.NET profile provider and membership provider. Because aExpense uses Windows authentication, it recognizes the credentials used when employees log on to the corporate network and doesn't need to prompt them again for their user names and passwords.

The aExpense access control rules use application-specific roles such as "Employee" and "Manager" that are accessed through the ASP.NET role management provider. Access control is intermixed with the application's business logic. It uses a simple SQL Server database for storing application data, and LINQ to SQL as its data access mechanism. The application is configured to connect to SQL Server by using integrated security, and the website uses a service account to log on to the database. The aExpense application also uses the Enterprise Library Logging Application Block and the Exception Handling Application Block for logging diagnostic information from the application.

Figure 1 shows a whiteboard diagram of the structure of the on-premises aExpense application.

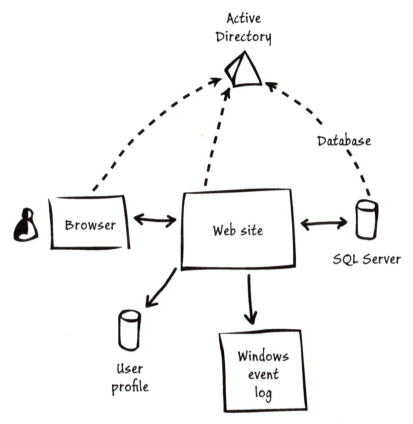

FIGURE 1
aExpense as an on-premises application

GOALS AND REQUIREMENTS

Adatum wants to explore the opportunities for cloud hosting the aExpense application in an attempt to maximize performance and availability, even during periods of peak usage, while minimizing the associated costs. The goals at this stage are to avoid, as far as possible, changes to the application code and the associated administrative functions while taking advantage of the flexibility and scalability offered by cloud hosting.

Therefore, as the first step in the migration path, Adatum has decided to deploy the aExpense application to the cloud using an IaaS approach. This will avoid any requirement to refactor the application or to make significant changes to the code because it can continue to run on a hosted server running Windows Server. However, the developers and administrators must still consider what, if any, changes are required to the application; and the impact of moving it from their on-premises datacenter to the cloud.

Adatum chose to use Windows Azure because of its wide range of capabilities for hosting both the application code and the data, and the availability of additional Windows Azure services that are appropriate to meet the application's requirements. For example, Adatum wants to continue to use Active Directory for authenticating users, and be able to integrate the application with its existing on-premises System Center Operations Manager.

Adatum also wants to be able to deploy the application in a secure and repeatable way to a staging environment first, and then to a production environment when testing is complete. After deployment, Adatum's administrators want to be able to scale the application to cope with varying usage patterns, monitor its execution, and be able to adjust configuration settings to fine tune it.

OVERVIEW OF THE SOLUTION

This section of the chapter explores the high-level options Adatum had for migrating the aExpense application during this step. It shows how Adatum chose an appropriate hosting mechanism for the application and for the data it uses, and how Adatum can establish connectivity between the cloud-hosted servers and its on-premises corporate network.

Options for Hosting the Application

Having decided on an IaaS approach for this initial step, Adatum must consider the hosting options available. Windows Azure provides the following features for IaaS deployment:

- **Virtual Machine.** This feature allows you to provision a virtual machine in the cloud with a choice of operating systems, and with a range of options for preinstalling a database server on the virtual machine. Alternatively, you can upload a prebuilt VM to the cloud. You can use it to run any software compatible with your chosen operating system, and configure the operating system and services as required. Virtual Machine instances maintain state between restarts, and so are suitable for use when software or services you install require state to be maintained.

- **VM Role.** This feature allows you to host your own customized instance of the Windows Server 2008 R2 Enterprise or Windows Server 2008 R2 Standard within a Windows Azure data center. However, the role does not save state when Windows Azure restarts or redeploys it as part of the operations executed automatically within the data center (such as when the role fails), and so it is not suitable for scenarios where the software or services you install require state to be maintained.

- **A set of associated services** that provide connectivity and additional functionality for IaaS applications. These services include Windows Azure Connect and Virtual Networks for providing connectivity to hosted servers, and functional services such as claims-based access control, Service Bus relay and messaging, database synchronization, and caching.

The major difference between Windows Azure Virtual Machines and the VM role is the behavior when the machine or role is reinitialized. This can occur following a hardware failure, or automatically as the Windows Azure internal systems manage allocation of resources by moving instances to a different physical server.

Any changes to a virtual machine, such as installing applications or configuring the operating system, are persisted when the instance is reinitialized – this is what is meant by the machine being able to maintain its state. VM role instances do not persist these changes. Any changes such as installing applications or configuring the operating system are lost and the role will return to the originally deployed configuration when reinitialized (although changes are persisted across restarts as long as the VM role is not reinitialized).

This means that you must redeploy the complete VM role image each time you need to make changes to it. When you use a virtual machine you do not need to do this. You can install and update applications on the virtual machine without needing to redeploy it every time, which makes it an attractive option for both testing and production deployment.

> *For more information about how Windows Azure manages maintaining state in virtual machines, see the section "Storing and Backing Up Virtual Machines" later in this chapter.*

After considering the available application hosting options, Adatum chose to host the application in a Windows Azure Virtual Machine at this stage of their migration process. Developers and testers will be able to deploy to virtual machine instances in exactly the same way as when using an on-premises server. Administrators and operators will be able to manage the live production server and deploy applications to it just as they do with on-premises servers.

Affinity Groups

When you first create namespaces and instances of Windows Azure services, you must specify the location or datacenter where they will be deployed. However, for some services you can specify only a region rather than a specific datacenter. Therefore, to ensure that elements of your application (such as the compute and storage services) are co-located in the same datacenter as close together as possible you specify an affinity group.

When you specify an affinity group, which must be done when creating the services (you cannot move services into a different affinity group after they are created), Windows Azure will attempt to locate all of the services in the group in the same cluster in the same datacenter. This will maximize performance, reduce latency, and eliminate unnecessary data transfer costs.

For information about using affinity groups, see "Importance of Windows Azure Affinity Groups."

Adatum will define an affinity group when it creates the virtual network that aExpense uses to access the on-premises Active Directory service. Adatum will also add all the virtual machines to the same affinity group when it creates them.

Availability Sets

In Windows Azure, fault domains are a physical unit of failure. Each virtual machine in an availability set is deployed to a different fault domain in Windows Azure. This helps to ensure that your application stays available during network failures, local disk hardware failures, and other outages.

However, improving the reliability and responsiveness of the aExpense application by deploying multiple copies of the virtual machines in an availability set will add to the running costs incurred by Adatum.

Options for Hosting the Data

The existing on-premises aExpense application stores its data in SQL Server. Therefore, Adatum also had to consider how to provide a comparable capability for the application when hosted in the cloud. Adatum has several options:

- **Keep the data on-premises.** In theory this is a simple to implement option. The cloud-hosted application would communicate with the on-premises database over the Internet. However, it raises several issues, such as the security of the connection and the requirement to expose the database server outside of Adatum's internal network. One approach to resolve this would be to use Windows Azure Connect or Virtual Networks to provide a private secure connection. Even with this approach, there are issues around the additional connection latency and the possibility of intermittent connectivity failures, which would require extensive caching in the application and a connection retry mechanism.

- **Deploy the data in a hosted SQL Server.** This approach would require Adatum to deploy a separate virtual machine to run SQL Server, although this can be easily provisioned using the templates available in the Windows Azure portal. Connection latency and intermittent connectivity would be minimized by deploying it in the same datacenter as the application.

- **Deploy the data in Windows Azure SQL Database**. This approach would require Adatum to subscribe to the managed data service offered by Windows Azure, and accept the few limitations that it imposes when compared to the full version of SQL Server. However, it is a viable and useful option that Adatum will consider in a future migration step. Chapter 4, "Moving to Windows Azure SQL Database," of this guide looks in detail at the differences between SQL Server and Windows Azure SQL Database, and the factors that affect Adatum's choice at that stage of the migration process.

- **Deploy the data in Windows Azure table and blob storage**. This approach would require Adatum to rewrite the data access code in the application to accommodate the differences between the relational, SQL-based data access approach and the less structured storage mechanisms used in Windows Azure storage. However, there are some specific advantages offered by Windows Azure storage such as lower cost and vast storage capability, and Adatum will consider this in a future migration step. At this stage Adatum's goal is to minimize the changes required to the code. Chapter 7, "Moving to Windows Azure Table Storage," of this guide looks in detail at the differences between relational databases and Windows Azure storage.

> You do not have to use Windows Server and SQL Server when you choose the IaaS approach for your database. You can install a range of operating systems and database servers in a hosted virtual machine using the templates available in the Windows Azure portal. For example, you may decide to use MySQL running on Linux as an alternative to SQL Server running on Windows Server.

- **Deploy the data in a custom store or repository**. You can deploy a range of operating systems and software on a hosted Virtual Server, and so you can continue to use you existing data stores and repositories. For example, if Adatum used a custom or third party data storage mechanism based on a non-standard file format, or even just use a simple file server, it could be deployed on Windows Server or Linux on a virtual machine. However, as Adatum uses SQL Server running on Windows Server in its on-premises datacenter, this option is not applicable to its migration strategy.

  > *Chapter 4, "Moving to Windows Azure SQL Database," describes the differences between using a hosted SQL Server and Windows Azure SQL Database, and how the developers at Adatum chose a data storage mechanism for the aExpense application in the subsequent steps of their migration path.*

After considering the available data storage options, Adatum chose to host the data in SQL Server running on a separate virtual machine in the same datacenter as the application.

Connectivity for Authentication

Adatum's aExpense application relies on connectivity to the Adatum corporate domain to authenticate users against Active Directory, and Adatum wants to maintain the existing mechanism when migrating the application to the cloud. This will avoid the need to make changes to the code while Adatum evaluates the results of the first step in its migration path. In future migration steps Adatum will consider changing the authentication approach to use another mechanism, such as claims and federated identity.

However, to continue to use Active Directory for authentication when the application resides in the cloud, Adatum must establish connectivity between the application and its on-premises corporate domain network. One of the advantages of Windows Azure as a hosting environment is that it includes services to enable connectivity that is safe, secure, and easy to set up.

Adatum has two options for establishing this connectivity between the cloud-hosted application and its corporate domain network:

Adatum must establish connectivity between its virtual machines in the cloud and its on-premises Active Directory server to continue to use Windows Authentication when it deploys to a virtual machine in the cloud. The current version of aExpense also relies on Windows Authentication to connect to SQL Server.

- **Windows Azure Connect**. This technology allows administrators to set up a direct connection between a cloud-hosted virtual machine (or a cloud services role) and an on-premises computer. It relies on endpoint software installed in the remote and on-premises computers, which establishes a secure connection across the Internet. Connect does not require ports to be opened in the corporate firewall, and will usually work when network address translation (NAT) routing is in use. It is easy to set up and manage, and provides a secure connection.

- **Windows Azure Virtual Networks**. This technology uses the virtual private network (VPN) approach to establish connectivity across computers and hosted services in Windows Azure and on-premises. Computers, cloud-hosted virtual machines, and cloud service roles are configured on the network in the same way as when running on premises, and a VPN router on the corporate network establishes connectivity between the network segments. Effectively, this extends the corporate network into the cloud, allowing existing services and applications to be used with no special software installation required on the on-premises computers. Virtual networks are ideally suited to more complex scenarios where scalability and additional control are required.

Virtual networks are typically used where you need to connect virtual machines and cloud service roles together, for connecting between the cloud and on-premises, or when you need a virtual machine or a role to maintain the same IP address when redeployed. The video presentation "Migrating Applications to Windows Azure Virtual Machines" contains a wealth of information about setting up virtual machines and connecting then using Windows Azure Virtual Networks.

Both of these options will allow Adatum to continue to use Active Directory for authentication. However, there are some limitations when using Windows Azure Connect. The Active Directory server must also host DNS for the domain (the Connect service cannot be used where a separate DNS server is used), and administrators will need to install the Connect service endpoint software on the Active Directory computer. If this is not an acceptable scenario, Adatum can configure the cloud-hosted virtual machine as part of a Windows Azure Virtual Network. Although this is more complex to configure, it removes the requirement for installing endpoint software.

An alternative approach Adatum could take would be to install Active Directory on a virtual machine running in Windows Azure, and connect it to their on-premises Active Directory. However, this means that Adatum must pay for the additional Windows Azure resources it uses. In future releases of Windows Azure, the Access Control service it provides will be extended to allow integration with an on-premises Active Directory.

Session Support and Load Balancing

Applications that run in an on-premises datacenter often rely on intrinsic features of the network that are not supported in the cloud, and this can affect the way that the application works when deployed to the cloud. A typical example related to the aExpense application is the way that Adatum implements an on-premises web farm by using Windows Network Load Balancing (NLB), and the impact this has on the way that it stores users' session data.

The on-premises aExpense web application uses the built-in ASP.NET in-memory session mechanism to support sessions for each user. This means that session data is only available on the server that initiates the session, and so users must be routed to the same server on each request during a session. Adatum configured affinity for NLB in their datacenter so that this occurs automatically.

For more information about NLB see "Overview of Network Load Balancing."

We could also use Windows Azure Connect to link the cloud-hosted application to an on-premises database by installing the Connect service endpoint software on the database server computer if we needed to keep that database on-premises. Unlike opening a public endpoint in SQL Server, the Connect service provides a secure channel between the application and the database. However, in many organizations installing this type of software on the database server may be frowned upon, and this may preclude you from using an on-premises database with your cloud-hosted applications.

However, affinity is not directly supported in Windows Azure (although it can be implemented with a plug-in or with custom code). If Adatum wants to run multiple instances of the virtual machine that hosts the eExpense application in Windows Azure, the developers must change the way session data is stored so that it is available to all instances. For example, they could switch to using the ASP.NET SQL Server session store provider, Windows Azure storage, or use Windows Azure Caching. Chapter 3 of this guide, "Moving to Windows Azure Cloud Services," explores these options and shows how Adatum's developers implemented session storage during the next stage of their migration strategy.

At the moment Adatum is most concerned with getting the application running in the cloud to explore the possibilities, validate usability and performance, and gauge user acceptance. Adatum wants to avoid making changes to the code. Therefore, in this step of its migration plan, Adatum will run only one instance of the virtual machine that hosts the aExpense application in the cloud.

> *For information about how to enable load balancing between multiple instances of Virtual Machines in Windows Azure see the section "Load balancing virtual machines" in the topic "Virtual Machines." The blog post "Setting up a webfarm using Windows Azure Virtual Machines" provides a walkthrough of the procedure.*

Integrating with On-Premises Services

The existing on-premises aExpense application integrates with other applications within Adatum's corporate network. For example, it reads and writes text files that are used to import and export expenses data for processing by Adatum's other on-premises systems. Therefore Adatum must consider how it can establish the equivalent functionality when the application is hosted in the cloud.

> *The Visual Studio solution **BeforeAzure** discussed in this chapter does not include this data import and export functionality. The Visual Studio solution **Azure-TableStorage** discussed in Chapter 7, "Moving to Windows Azure Table Storage," shows how Adatum added the data export feature to the aExpense application.*

When using IaaS and virtual servers in the cloud, the application can write to the virtual disk in the same way as when running on a physical server. Administrators can enable direct connectivity with the machine by using Windows Azure Connect or a Windows Azure Virtual Network to allow access to disk files stored on the drives of the remote server.

Alternatively, Adatum's developers could change the code so that it serializes the files and stores them in either Windows Azure storage or in the database. As the files will be read and generated remotely, and streamed to and from the in-house application over the Internet, this is not an issue because they will need to be serialized anyway and can be reconstructed on the on-premises server.

Both of the connectivity options, Windows Azure Connect and Windows Azure Virtual Network, allow Adatum to connect the cloud-hosted virtual machines running the application and SQL Server to their corporate domain. This means that existing tools and practices will continue to work without changes being required. For example, administrators can browse the remote machines, use database management tools, run existing scripts, and use the same monitoring and management applications. For more details of how Adatum's administrators will manage the remote servers and application, see the section "Management and Monitoring" later in this chapter.

A virtual machine or a VM role or can be used simply as a file server in the cloud. When connected through Windows Azure Connect or Windows Azure Virtual Networks it will be accessible from on-premises computers and by other applications running in the cloud.

INSIDE THE IMPLEMENTATION

Now is a good time to walk through the process of migrating aExpense into a cloud-based application in more detail. As you go through this section, you may want to download the Microsoft Visual Studio development system solution from *http://wag.codeplex.com/*. This solution contains an implementation of the aExpense application (in the **BeforeAzure** folder) as it is when running on-premises. If you are not interested in the mechanics, you should skip to the next section.

Profile Data

Before the migration, aExpense used the ASP.NET profile feature to store application-specific user settings. Adatum tries to avoid customizing the schema in Active Directory, so aExpense stores a user's preferred reimbursement method by using the profile feature. The default profile provider stores the profile properties in a database.

Using the profile feature makes it very easy for the application to store small amounts of user data. Adatum enabled the profile feature and specified the properties to store in the Web.config file.

We don't like to customize the Active Directory schema if we can possibly avoid it. Schema changes have far-reaching implications and are difficult to undo.

```xml
XML
<profile defaultProvider="SqlProvider">
  <providers>
    <clear />
    <add name="SqlProvider"
         type="System.Web.Profile.SqlProfileProvider"
         connectionStringName="aExpense"
         applicationName="aExpense" />
  </providers>
  <properties>
    <add name="PreferredReimbursementMethod" />
  </properties>
</profile>
```

Then developers can access a profile property value in code like this.

```csharp
C#
var profile = ProfileBase.Create(userName);
string prm =
    profile.GetProperty<string>("PreferredReimbursementMethod");
```

After migration, aExpense continues to use the profile system to store the preferred reimbursement method for each user.

The application also uses ASP.NET membership to store the users preferred name, and ASP.NET role management for the custom roles used in the aExpense application. You can see how these are configured in the Web.config file, and how they are accessed in the **User-Repository class**.

Connecting to the Hosted SQL Server

Connecting to a hosted SQL Server running on a virtual machine instead of a SQL Server on-premises requires only a configuration change.

Before the migration aExpense stored application data in a SQL Server database running on-premises. In this first phase, the team moved the database to a hosted SQL Server running on a virtual machine. The data access code in the application remains unchanged. The only thing that needs to change is the connection string in the Web.config file.

```XML
<add name="aExpense" connectionString=
  "Data Source={Server Name};
   Initial Catalog=aExpense;
   Integrated Security=True;
   Encrypt=yes;"
  providerName="System.Data.SqlClient" />
```

This connection string uses the same SQL Server security mode as the on-premises solution. To continue using Windows Authentication mode, both the virtual machines must be able to access the Adatum on-premises Active Directory service. The server name must be set to the server name of the virtual machine where SQL Server is hosted. Adatum also chose to encrypt the connection to SQL Server; for more information see *"Encrypting Connections to SQL Server "* on MSDN.

> *If you choose to use SQL Server authentication instead of Windows authentication to connect to SQL Server, your connection string will include a user name and password in plain text. In this scenario, you should consider encrypting your SQL connection string in the Web.config file; to do this, you can use the Pkcs12 Protected Configuration Provider. For more information, see Chapter 4, "Moving to Windows Azure SQL Database."*

The connection string also specifies that all communications with Windows Azure SQL Database are encrypted. Even though your application may reside on a computer in the same data center as the server hosting SQL Server, you should treat that connection as if it was using the internet.

Database Connection Timeouts and Dropped Connections

If a connection to SQL Server times out or drops while your application is using the connection, you should immediately try to re-establish the connection. If possible, you should then retry the operation that was in progress before the connection dropped, or in the case of a transaction, retry the transaction. It is possible for a connection to fail between sending a message to commit a transaction and receiving a message that reports the outcome of the transaction. In this circumstance, you must have some way of checking whether the transaction completed successfully in order to determine whether you must retry it.

You can also add protection to your hosted SQL Server by configuring the Windows Firewall on the virtual machine so that it allows only a specific range of permitted IP addresses to connect to your database server. However, this is complicated by the fact that you may not always be able to determine ahead of time the IP address of your virtual machine instance(s). For example, when you redeploy a virtual machine, the IP address may change.

When using a hosted SQL Server, you can continue to use the same techniques as you do in on-premises applications. As long as the virtual machines hosting the application and the SQL Server are co-located in the same datacenter, connection failures should arise no more often than when running in your own datacenter. However, it is useful to consider implementing a mechanism that can alleviate issues with failed connections. In Chapter 4, "Moving to Windows Azure SQL Database," of this guide you will see how Adatum chose to use the Enterprise Library Transient Fault Handling Block to manage connection retries when connecting to the database.

Deploying the aExpense Application to Windows Azure Virtual Machines

This section outlines the steps that Adatum took to deploy the aExpense application to two Windows Azure virtual machines: one for the web application and one for the SQL Server database. For a more detailed description of the steps, see the Hands-on Labs that are available for this guide.

Deploying the Web Application

By configuring appropriate endpoints for your virtual machine in the Windows Azure portal, you can improve security by limiting its publicly available TCP ports.

1. Create a virtual network to enable connectivity to the Adatum on-premises Active Directory server.

2. Create a virtual machine based on the Windows Server image available in the portal and ensure that the virtual machine is part of the virtual network.

3. Set up remote desktop connectivity to enable access to the virtual machine from on-premises.

4. Join the machine to the Adatum Active Directory.

5. Install .NET 4 on the virtual machine.

6. Install and configure the dependencies of the aExpense application: install Internet Information Server (IIS), enable Windows Communication Foundation in IIS, enable Application Pool User Profile in IIS, install certificates required for HTTPS, and configure HTTPS in IIS.

7. Copy the application code to the server, and add a web application in IIS.

Deploying the Database

1. Create a virtual machine that includes SQL Server. Adatum did this by using an image available in the portal that includes SQL Server 2012, but it could have installed a different version. This virtual machine should be part of the same virtual network and affinity group as the virtual machine that hosts the web application.

2. Set up remote desktop connectivity to enable access to the virtual machine from on-premises.

3. Join the machine to the Adatum Active Directory.

4. Open port 1433 in Windows Firewall to allow access to the database (you may want to use a different port to help hide SQL Server), and enable network connectivity in SQL Server for the TCP/IP protocols.

5. Migrate the on-premises database used by the aExpense application to the SQL Server installation in the virtual machine. Ensure that the domain account used by the application pool on the web server has access to the database.

Adatum is using a virtual network to enable connectivity to the on-premises services that the aExpense application uses. The virtual network is not needed to enable connectivity between the two Windows Azure virtual machines.

During testing the development team at Adatum is working with sample data. However, the developers must consider how they will deploy both test and live data to the SQL Server instances running in the hosted virtual machines. To migrate an existing database schema and data to a hosted SQL Server, Adatum can use SQL Server Management Studio. The SQL Server Integration Service and other export tools it contains allow administrators to connect to both the local and the Windows Azure hosted servers to create a copy of the database and its data in the hosted server.

An alternative is to export the schema—and, if required, the data—as a Transact-SQL script, and then connect to the remote server using Remote Desktop Protocol (RDP) and run the script using SQL Server Management Studio or the command line on that server. Using a script allows Adatum to include this as part of the build process for the test server running in Windows Azure staging, so that the application can be tested without using the live data.

TESTING, DEPLOYMENT, MANAGEMENT, AND MONITORING

This section discusses application life cycle management topics for applications hosted in Windows Azure. Although specific requirements will vary between applications and across organizations, everyone develops applications, then tests them, and finally deploys them. Typical factors that must be considered are where applications should be tested, and how the deployment process should be managed to make sure that only authorized personnel have access to the live production environment.

In this chapter the focus is on the specific scenarios related to IaaS deployment using hosted virtual machines. In subsequent chapters you will find additional information related to the subsequent stages of the migration process Adatum followed. If you choose to use the Windows Azure VM role instead of virtual machines, you will need to establish a testing and deployment procedure that encompasses redeploying the server images each time you change the application or reconfigure the server.

We want to have a clearly defined process for deploying applications to Windows Azure that allows us to control and trace operations carried out on our subscriptions.

Adatum has a well-defined set of existing processes for deploying applications to its on-premises servers. The development and testing teams use separate servers for testing, staging, and production. When the development team releases a new version of the application, its progression through the testing, staging, and production environments is tightly controlled. Very rarely, though, small fixes, such as updating the text on an ASPX page, are applied directly to the production servers.

Adatum has a number of goals for application life cycle management for Windows Azure. Adatum wants to maintain its clearly defined process for deploying applications to Windows Azure, with clearly defined roles and responsibilities. More specifically, it wants to make sure that access to the live environment is only available to a few key people, and that any changes to the live environment are traceable.

You can use the Windows Azure API to get a list of create, update, and delete operations that were performed on a subscription during the specified timeframe. For more information, see "List Subscription Operations."

In addition, Adatum wants to be able to roll back the live environment to a previous version if things go wrong. In a worst-case scenario, they would like to be able to pull the application back to be an on-premises application.

Testing Applications for Virtual Machine Deployment

One of the major advantages of using virtual machines to test and deploy applications is that the application lifecycle is identical to that when using your own on-premises datacenter. Nothing needs to change; all of your existing procedures and practices continue to work in the same way. The only difference is that the servers are no longer in your own datacenter, but located in a Windows Azure datacenter. Other than the increase in network latency and throughput compared to a local network, developers, testers, and operators should be unaware of any differences.

Developers can run unit tests and ad-hoc tests on their local computers. The on-premises build server also runs a suite of tests as a part of the standard build process. The application can take its data from a local server running a test copy of the SQL Server database for the aExpense application. This is no different from the normal development practices for on-premises applications.

For final testing to confirm correct operation, the application is deployed to a test environment consisting of one or more Windows Azure virtual machines. While the runtime environment in a virtual machine is identical to a local Windows Server installation, testing in Windows Azure ensures that the application works correctly in an environment that uses the same network and infrastructure as the live Windows Azure runtime environment. Testers can change the connection string in the application so that it uses data stored in a SQL Server running on a virtual machine in Windows Azure.

Test and Production Environments

Adatum wants to continue to be able to deploy an application to either a staging or a production area. This is a typical scenario where new versions of applications are deployed first to the staging area and then, at the appropriate time, deployed to the production area.

When using virtual machines and a virtual network, Adatum can provide test and production (live) environments simply by configuring network subnets, segments, and permissions so that only specific people can access each environment; just as it does in the existing on-premises test and production environments.

You should try to ensure that your local Windows Server testing environment is configured identically to your Windows Server virtual machines in Windows Azure, and has the same patches installed.

Because development and testing staff don't have access to the production environment, there's no risk of accidentally deploying to the live environment.

The test and production applications use Windows authentication to connect to SQL Server. The test application connects to a test database server, while the production version uses a different server name in the connection string to connect to the live database in the production environment. The application pool in IIS runs under a different identity for test and production areas.

Figure 2 summarizes the application life cycle management approach at Adatum. It uses a single virtual machine for the test environment, which connects to a virtual machine hosted SQL Server containing the test database. The production environment also uses a single virtual machine, which connects to a virtual machine hosted SQL Server containing the live database.

By configuring the endpoints in the testing subscription to use non-standard ports you can reduce the chances that casual users will access the test version of the application.

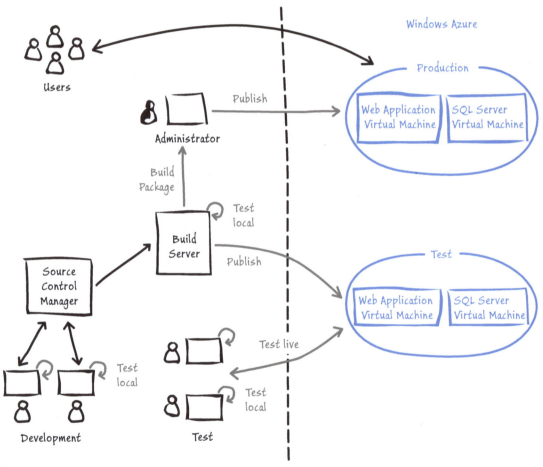

FIGURE 2
Adatum's application life cycle management environment

Even though there is a separation between on-premises and cloud-hosted environments, indicated by the dashed line in Figure 2, keep in mind that this is just a logical division. The Windows Azure virtual network that connects all the computers hides the division so that the virtual machines appear to users and to automated software processes running on-premises as though they are part of the on-premises network.

> You can deploy an application to your Windows Azure test environment just while you run the tests, but don't forget that any time you do something in Windows Azure–even if it's only testing–you will be charged for the virtual machines and for data transferred out of the datacenter. You can minimize data transfer costs by locating all your test virtual server instances in the same datacenter. However, when you use Windows Azure Virtual Machine, you are billed even if the virtual machine is shut down because Windows Azure reserves the resource for you. You should use PowerShell scripts to import and export your virtual machine settings, and delete the test virtual machines when you are not using them.

Management and Monitoring

Administrators at Adatum can connect to the virtual machines running the application and the SQL Server using RDP, and manage them in exactly the same way as when the application and database were installed in their local server room. All of the built-in Windows Server monitoring and management facilities are available, and operate in the same way as when on-premises. The only likely difference will be additional latency through the use of the public Internet as the connectivity transport.

The on-premises version of aExpense uses the Logging Application Block and the Exception Handling Application Block to capture information from the application and write it to the Windows event log. Administrators can view these logs to obtain information about the application and the operating system. They can also use tools such as Windows Performance Counters and Task Manager.

In Adatum's case, automated monitoring of on-premises applications takes place using System Center Operations Manager. By establishing connectivity between the hosted virtual machines and the on-premises Operations Manager server, Adatum's administrators can continue to monitor the hosted servers in exactly the same way as when they were running on-premises.

You can think of your virtual machines as rehydrated instances of the VHDs stored in Windows Azure blob storage. It's easy to initialize a virtual machine from the stored VHD or capture a virtual machine to a VHD in much the same way as you would when using an on-premises hypervisor such as Windows Hyper-V.

We use scripts because it means that we can reliably repeat operations. We're also planning to enhance some of our scripts to record when they were run, by whom, and what parameters were used. You will see more details of how we use scripts in Chapter 3, "Moving to Windows Azure Cloud Services," of this guide.

Storing and Backing Up Virtual Machines

When you create a virtual machine, the virtual hard disk (VHD) image is held in blob storage and is used to initialize the virtual machine. You can create a virtual machine from one of the standard templates available in the Windows Azure Management Portal, in which case a copy of that template image is placed in a blob in your Windows Azure storage account. Alternatively, you can create the virtual machine by uploading your own VHD into blob storage.

Once the VHD is in blob storage, you can initialize one or more virtual machines from it. When the virtual machine is running, any changes are replicated in the blob so that they are persisted when the machine restarts. Blobs are automatically replicated three times for durability, and you can copy the blob to another location in your storage account (or to another storage account) to create a backup. You can also capture an image of a virtual machine and place it in blob storage.

> For more information about how Windows Azure stores and manages VHDs and virtual machines, see "Data Series: Exploring Windows Azure Drives, Disks, and Images."

Managing Virtual Machine Instances

Adatum will need to manage the virtual machine instances it uses to perform tasks such as stopping, starting, adding, and deleting instances, and for managing blob storage that holds the VHD images. For example, the instances used for testing the application are charged even when not running, and so Adatum will need to remove these when not in use. They can easily be reinitialized from the VHD image held in Windows Azure blob storage when required.

There are three ways to access a Windows Azure environment to perform management tasks such as this. The first is through the Windows Azure Management Portal, where a single Microsoft account has access to everything in the portal. The second is by using the Windows Azure Service Management API, where API certificates are used to access to all the functionality exposed by the API. The third is to use the Windows Azure Management or Node.js Power-Shell cmdlets. Within Adatum, almost all operations that affect the test or production environment are performed using scripts based on the PowerShell cmdlets instead of the Management Portal.

Elasticity and Scaling

In addition to managing and monitoring the application, Adatum wants to be able to scale the application to meet the additional load it typically encounters during the last few days of each month. At present, Adatum has deployed only a single instance of the application and the database server. However, during the busy period administrators can use the Windows Azure Management Portal or scripts to change the number of instances deployed.

> *As described earlier in this chapter, the on-premises version of the aExpense application uses in-memory session state management. Adatum recognizes that this limits the ability of the application to scale out across multiple virtual machines until it replaces the session state management with a web farm friendly solution, such as storing session state in the SQL Server database, in Windows Azure table storage, or in the Windows Azure caching service. We discuss these options in more detail in Chapter 3, "Moving to Windows Azure Cloud Services."*

The options available to Adatum for scaling the aExpense application are to use a larger virtual machine at times when demand is high; however, this would require Adatum to briefly shut down the application as it moved the application from one virtual machine to another. Alternatively, if Adatum modifies the session state management to be web farm friendly, it can scale the application out by adding additional virtual machines when demand is high. For example, they may decide from the results of monitoring the performance and response times that an additional two instances of the virtual machine hosting the application should be available for the last three days of the month, and then scale back to a single instance on the first day of the following month.

At the time this guide was written, the additional costs for running two extra small virtual machine instances for three days is less than $20.00, and so for a total cost of under $ 250.00 per year Adatum can make the aExpense application much more responsive during the busy month-end period.

Isolating Active Directory

The aExpense application uses Windows Authentication. Because developers do not control the identities in their company's enterprise directory, it is sometimes useful to swap out Active Directory with a stub during the development of your application.

When you scale out using virtual machines in Windows Azure, you must be careful to delete the virtual machines when you scale in again. You will be charged for virtual machines even if they are shut down in Windows Azure.

The on-premises aExpense application (before the migration) shows an example of this. To use this technique, you need to make a small change to the Web.config file to swap Windows Authentication for Forms Authentication and then add a simulated LDAP profile store to the application. Swap Windows Authentication for Forms Authentication with the following change to the Web.config file.

```xml
XML
<authentication mode="Forms">
  <forms name=".ASPXAUTH"
      loginUrl="~/SimulatedWindowsAuthentication.aspx"
      defaultUrl="~/default.aspx" requireSSL="true">
  </forms>
</authentication>
```

You need to add a logon page to the application that enables you to select the user that you want to use for testing. In aExpense, this page is known as SimulatedWindowsAuthentication.aspx.

You also need to add a class that simulates an LDAP lookup for the Active Directory attributes that your application needs. In this example, the **GetAttributes** method simulates the LDAP query "&(objectCategory=person)(objectClass=user);costCenter;manager;displayName".

```csharp
C#
public static class SimulatedLdapProfileStore
{
    public static Dictionary<string, string> GetAttributesFor(
        string userName, string[] attributes)
    {
        Dictionary<string, string> results;

        switch (userName)
        {
            case "ADATUM\\johndoe":
                results = new Dictionary<string, string>
                {
                    { "costCenter", "31023" },
                    { "manager", "ADATUM\\mary" },
                    { "displayName", "John Doe" }
                };
                break;

            ...

        }

        return results;
    }
}
```

*These code samples come from the **BeforeAzure** solution in the downloadable solutions.*

MORE INFORMATION

All links in this book are accessible from the book's online bibliography available at: *http://msdn.microsoft.com/en-us/library/ff803373.aspx.*

MSDN is a good starting point to learn more about deploying applications to *Windows Azure.*

The *Windows Azure Manage Center* contains links to plenty of resources to help you learn about managing applications in Windows Azure and about *Windows Azure Virtual Machines.*

You can find out more about networks and connecting cloud hosted resources to on-premises resources in the *Networks* topic on MSDN.

For a detailed guide to creating a Windows Azure Virtual Machine, see *"Create a Virtual Machine Running Windows Server 2008 R2."*

For a detailed guide to installing SQL Server on a Virtual Machine, see *"Provisioning a SQL Server Virtual Machine on Windows Azure."*

The Hands-on Labs that are available for this guide include a step-by-step guide to setting up Virtual Machines, SQL Server, and Windows Active Directory. For more details, see the *"Windows Azure Guidance"* site on CodePlex.

3 Moving to Windows Azure Cloud Services

This chapter walks you through the second step Adatum took in migrating their aExpense application to Windows Azure. You'll see an example of how to take an existing business application, developed using ASP.NET, and adapt it to run in the cloud using Windows Azure Cloud Services.

THE PREMISE

At the end of the first migration step, Adatum had a version of the aExpense application that ran in the cloud using the IaaS approach. When the team at Adatum developed this version, they kept as much of the original application as possible, changing just what was necessary to make it work in Windows Azure.

During this step in the migration, Adatum wants to evaluate changing to use a PaaS approach instead of using the IaaS approach with virtual machines. Adatum also wants to address some issues not tackled in the previous deployment. For example, Adatum's developers are investigating adapting the aExpense application to use claims-based authentication instead of Active Directory, which will remove the requirement to connect back to the corporate domain through Windows Azure Connect or Virtual Networks.

Adatum also needs to consider other issues such as how to handle session state when there may be multiple instances of the application running, and how to deploy and use configuration settings required by the application.

GOALS AND REQUIREMENTS

In this phase, Adatum has a number of goals for the migration of the aExpense application to use a PaaS approach. Further optimization of the application for the cloud, and exploiting additional features of Windows Azure, will come later.

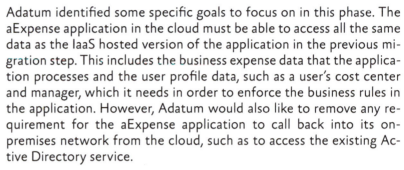

Adatum identified some specific goals to focus on in this phase. The aExpense application in the cloud must be able to access all the same data as the IaaS hosted version of the application in the previous migration step. This includes the business expense data that the application processes and the user profile data, such as a user's cost center and manager, which it needs in order to enforce the business rules in the application. However, Adatum would also like to remove any requirement for the aExpense application to call back into its on-premises network from the cloud, such as to access the existing Active Directory service.

A second goal is to make sure that operators and administrators have access to the same diagnostic information from the cloud-based version of aExpense as they have from the on-premises and IaaS hosted versions of the application.

The third goal is to continue to automate the deployment process to Windows Azure. As the project moves forward, Adatum wants to be able to deploy versions of aExpense to Windows Azure without needing to manually edit the configuration files, or use the Windows Azure Management Portal. This will make deploying to Windows Azure less error-prone, and easier to perform in an automated build environment.

A significant concern that Adatum has about a cloud-based solution is security, so a fourth goal is to continue to control access to the aExpense application based on identities administered from within Adatum, and to enable users to access the application by using their existing credentials. Adatum does not want the overhead of managing additional security systems for its cloud-based applications.

Adatum also wants to ensure that the aExpense application is scalable so that when demand for the application is high at the end of each month, it can easily scale out the application.

Overall, the goals of this phase are to migrate aExpense to use a PaaS approach while preserving the user experience and the manageability of the application, and to make as few changes as possible to the existing application.

We want to avoid having to make any calls back into Adatum from the cloud application. This adds significantly to the complexity of the solution.

It would be great if we could continue to use tried and tested code in the cloud version of the application.

OVERVIEW OF THE SOLUTION

This section of the chapter explores the high-level options Adatum had for migrating the aExpense application during this step. It shows how Adatum evaluated moving from an IaaS to a PaaS hosting approach, and how it chose an appropriate mechanism for hosting the application in Windows Azure.

Evaluating the PaaS Approach for Hosting the Application

In the first migration step Adatum's goal was to avoid having to make any changes to the code so that it could quickly get the aExpense application running in the cloud, evaluate performance and usability, and gauge user acceptance. However, Adatum is also considering whether the IaaS approach they had followed was the best choice for their specific scenario, or whether a PaaS approach would be more cost effective in the long run while still providing all of the capabilities for performance, availability, manageability, and future development.

Adatum carried out some preliminary costing analysis to discover the likely differences in runtime costs for the aExpense application deployed using the IaaS model described in Chapter 2, "Getting to the Cloud," and using the PaaS model described in this chapter.

For a comparison of the options and features in Windows Azure for IaaS and PaaS, see the *"Windows Azure Features Overview."*

- In both the IaaS and PaaS deployments, the costs of data storage and data transfer will be the same: both use SQL Server, and both will transfer the same volumes of data into and out of Windows Azure.

- At the time of writing, the cost of running a medium sized virtual machine in Windows Azure is almost the same as the cost of running a Cloud Services role instance.

- The IaaS deployment uses a virtual network to provide connectivity with the on-premises Active Directory. At the time of writing, this costs approximately $ 37.00 per month.

> *Adatum used the Windows Azure pricing calculator, together with some calculations in an Excel spreadsheet. For more information, see Chapter 6, "Evaluating Cloud Hosting Costs." The Hands-on Labs associated with this guide include a cost calculation spreadsheet and describe how you can calculate the approximate costs of running your applications in Windows Azure.*

The hourly compute costs for Virtual Machines, Cloud Services roles, and Windows Azure Web Sites Reserved instances (when all are generally available at the end of the discounted trial period) are almost the same, and so the decision on which to choose should be based on application requirements rather than focusing on just the compute cost.

Although the PaaS and IaaS deployment models for the aExpense application are likely to incur similar running costs, Adatum also considered the saving it can make in administrative and maintenance cost by adopting the PaaS approach. Using a PaaS hosting model is attractive because it delegates the responsibility for managing both the hardware and the operating system to the partner (Microsoft in this case), reducing the pressure on Adatum's in-house administration staff and thereby lowering the related costs and enabling them to focus on business critical issues.

Options for Hosting the Application

Having decided on a PaaS approach, Adatum must consider the hosting options available. Windows Azure provides the following features for PaaS deployment:

- **Web Sites**. This feature provides the simplest and quickest model for deploying websites and web applications to Windows Azure. The cost-effective Shared mode (and the Free mode available at the time of writing) deploy multiple sites and applications for different Windows Azure customers to each instance of IIS, meaning that there can be some throttling of bandwidth and availability. Alternatively, at increased cost, sites and applications can be configured in Reserved mode to avoid sharing the IIS instance with other Windows Azure customers. A wide range of development languages and deployment methods can be used; and sites and applications can be progressively updated rather than requiring a full redeployment every time. Windows Azure Web Sites can also be automatically provisioned with a wide range of ecommerce, CMA, blog, and forum applications preinstalled.

- **Cloud Services**. This feature is designed for applications consisting of one or more hosted roles running within the Windows Azure data centers. Typically there will be at least one web role that is exposed for access by users of the application. The application may contain additional roles, including worker roles that are typically used to perform background processing and support tasks for web roles. Cloud Services provides more control and improved access to service instances than the Windows Azure Web Sites feature, with a cost for each role approximately the same as when using Web Sites Reserved mode. Applications can be staged for final testing before release.

- **A set of associated services** that provide additional functionality for PaaS applications. These services include access control, Service Bus relay and messaging, database synchronization, caching, and more.

The MSDN article "Windows Azure Websites, Cloud Services, and VMs: When to use which?" contains information about choosing a hosting option for your applications.

Choosing Between Web Sites and Cloud Services

Adatum considered the two Windows Azure PaaS approaches of using the Web Sites feature and the Cloud Services feature.

The Shared mode for Web Sites offers a hosting model that provides a low cost solution for deploying web applications. However, in enterprise or commercial scenarios the bandwidth limitations due to the shared nature of the deployment may mean that this approach is more suited to proof of concept, development, trials, and testing rather than for business-critical applications. Web Sites can be configured in Reserved mode to remove the bandwidth limitation, although the running cost is then very similar to that of Cloud Services roles. However, different sizes of Reserved mode instance are available and several websites can be deployed to each instance to minimize running costs.

The developers at Adatum realized that Reserved mode Web Sites would provide a useful platform for Adatum's websites that are less dependent on performing application-related functions. For example, Adatum plans to deploy its corporate identity websites and portals to the cloud in the future, as well as implementing additional gateways to services for mobile devices and a wider spectrum of users. Windows Azure Web Sites will be a good choice for these.

Windows Azure Web Sites can access a database for storing and retrieving data; however, unlike Cloud Services, they do not support the use of dedicated separate background processing role instances. It is possible to simulate this by using separate website instances or asynchronous tasks within a website instance but, as Adatum will require aExpense to carry out quite a lot of background processing tasks, the Cloud Services model that offers individually scalable background roles is a better match to its requirements.

You can also adopt a mixed approach for background processing by deploying the website in Windows Azure Web Sites and one or more separate Cloud Services worker roles. The website and worker roles can communicate using Windows Azure storage queues, or another mechanism such as Service Bus messaging.

Cloud Services make it easy to deploy applications that run on the .NET Framework, and it is possible (through not straightforward) to use other languages. In contrast, Windows Azure Web Sites directly supports a wide range of development languages such as node.js, PHP, Python, as well as applications built using the .NET Framework. However, the aExpense application already runs on the .NET Framework and so it will be easy to adapt it to run in a Windows Azure Cloud Service web role. It will also be possible to add background tasks by deploying one or more Cloud Service worker role instances as and when required.

Windows Azure Web Sites allows developers to use any tools to develop applications, and also supports a wide range of simple deployment and continuous automated update options that includes using FTP, Codeplex, Git, and Microsoft Team Foundation Server (TFS) as well as directly from Microsoft Web Matrix and Visual Studio. This would be a useful capability, especially for fast initial deployment without requiring any refactoring of the code and for deploying minor code and UI updates.

This wide choice of deployment and update capabilities are not available for Cloud Services where, other than changes to the configuration files, only a full deployment is possible. In addition, deployment to Cloud Services roles means that developers at Adatum will typically need to use Visual Studio; though this is already their default working environment.

The requirement of Cloud Services to deploy complete packages rather than individual files could be a disadvantage. However, Adatum wants to be able to deploy to a dedicated staging instance and complete a full application testing cycle before going live, and control the releases through versioning of each complete deployment rather than modifying it by using incremental updates. This means that the Cloud Services model is better suited to Adatum's needs. Developers and administrators can use scripts or custom applications that interact with the Windows Azure Management API to deploy and manage the application and individual role instances when using Cloud Services.

> It is possible to configure Web Sites to provide a staging area by using a separate instance, and integrate this with automated deployment through Codeplex or GitHub by configuring separate branches for each site in your source code repository.

Adatum would also like to introduce automatic scaling to the aExpense application running in the cloud, scaling out at the end of the month to handle increased demand and then scaling in again. The developers want to use the *Autoscaling Application Block* from the patterns & practices group at Microsoft, which can enable autoscaling for role instances under specified conditions or based on a predefined schedule, and it can be used only for Cloud Services roles.

Finally, Cloud Services allows Adatum to configure the firewall and endpoints for each role deployed to Windows Azure, and to configure virtual network connections between roles and on-premises networks. The ability to configure the firewall makes it easier to control public access to the website and other roles by defining endpoints and opening ports. The ability to use virtual networks makes it easy to interact with the roles from on-premises management tools, and use other integration services.

After initial investigation, and after considering the advantages and limitations of each approach, Adatum chose to use Cloud Services roles rather than Web Sites.

> *The Hands-on Labs available for this guide include an exercise that shows in detail how we deployed the application to Windows Azure Web Sites during this spike so that you can explore the use of Web Sites yourself.*

Service Instances, Fault Domains, and Update Domains

Adatum plans to deploy multiple instances of the aExpense application as a way to scale the application out to meet increased demand. It's easy for Adatum to add or remove instances as and when it needs them, either through the Windows Azure Management Portal or by using PowerShell scripts, so it only pays for the number it actually need at any particular time. Adatum can also use multiple role instances to enable fault tolerant behavior in the aExpense application.

> *Use multiple instances to scale out your application, and to add fault tolerance. You can also automate the deployment and removal of additional instances based on demand using a framework such as the **Enterprise Library Autoscaling Application Block.***

In Windows Azure, fault domains are a physical unit of failure. If you have two or more instances, Windows Azure will allocate them to multiple fault domains so that if one fault domain fails there will still be running instances of your application. Windows Azure automatically determines how many fault domains your application uses.

To handle updates to your application if you have two or more instances of a role, Windows Azure organizes them into virtual groupings known as update domains. When you perform an in-place update of your application, Windows Azure updates a single domain at a time; this ensures that the application remains available throughout the process. Windows Azure stops, updates, and restarts all the instances in the update domain before moving on to the next one.

> *You can also specify how many update domains your application should have in the service configuration file.*

Windows Azure also ensures update domains and fault domains are orthogonal, so that the instances in an update domain are spread across different fault domains. For more information about updating Windows Azure applications and using update domains, see *"Overview of Updating a Windows Azure Service."*

Before making this decision, we did create a spike to test the use of Windows Azure Web Sites, and to evaluate the performance and discover the capabilities. Note that, at the time of writing, Web Sites was still a preview release. The features of final version may differ from those described in this guide.

Options for Authentication and Access Control

Adatum wants to remove the requirement for the application to connect back to Adatum's on-premises Active Directory for authentication. In the IaaS version of the aExpense application described in the previous chapter, Adatum experimented with Windows Azure Connect and Windows Azure Virtual Network to connect the cloud-hosted virtual machines to its on-premises servers.

While this approach can still be implemented when hosting the application using Cloud Services, Adatum decided to investigate other options. These option include:

- **Run an Active Directory server in the cloud**. Adatum could provision and deploy a virtual machine in the cloud that runs Active Directory, and connect this to their on-premises Active Directory. However, this adds hosting cost and complexity, and will require additional management and maintenance. It also means that Adatum must still establish connectivity between the cloud-hosted and on-premises Active Directory servers.

- **Use standard ASP.NET authentication**. Adatum could convert the application to use ASP.NET authentication and deploy the ASPNETDB database containing user information to the hosted SQL Server instance that holds the application data. However, this means that users would need to have accounts in the ASP.NET authentication mechanism, and so it would not provide the same seamless sign-in as the existing application does through Windows Authentication against Active Directory.

- **Use a claims-based authentication mechanism**. Claims-based authentication is a contemporary solution for applications that must support federated identity and single sign-on. Users are authenticated by an identity provider, which issues encrypted tokens containing claims about the user (such as the user identifier, email address, and perhaps additional information). The application uses these claims to identify each user and allow access to the application. The advantage of this option is that users will continue to enjoy a single sign-on experience using their active directory credentials.

    In a future release, Windows Azure Access Control will be renamed to Windows Azure Active Directory (WAAD), and will expose functionality to support Windows Authentication in cloud-hosted applications. This would simplify implementation of authentication for Adatum.

The first alternative Adatum considered was to host Windows Active Directory in the cloud and continue to use the same authentication approach as in the on-premises version of the application. Active Directory can be hosted on a virtual machine and connected to an on-premises Active Directory domain through Windows Azure Connect or by using Windows Azure Virtual Networks.

Adatum will obviously still need to maintain an on-premises Active Directory domain for their internal applications, but the cloud-hosted domain controller will be able to replicate with the on-premises domain controllers. However, this means that Adatum would need to manage and pay for a virtual machine instance running all of the time just for authenticating users of the aExpense application. It would probably only make sense if Adatum planned to deploy a large number of applications to the cloud that use the cloud-hosted Active Directory server.

The second alternative Adatum considered was to use ASP.NET authentication. The developers would need to modify the code and add user information to the authentication database in the cloud-hosted SQL Server. If the application already used this mechanism, then this approach would remove any requirement to adapt the application code other than changing the authentication database connection string. However, as Adatum uses Windows Authentication in the application, this option was not considered to be an ideal solution because users would need a separate set of credentials to access the aExpense application.

The third alternative Adatum considered was to use claims-based authentication, and it has several advantages over the other two approaches. Adatum will configure an on-premises Active Directory Federation Services (ADFS) claims issuer in their data center. When a user tries to access the aExpense application in the cloud, that user will be redirected to this claims issuer. If the user has not already logged on to the Adatum domain, the user will provide his or her Windows credentials and the claims issuer will generate a token that contains a set of claims obtained from Active Directory. These claims will include the user's role membership, cost center, and manager.

> By moving to claims-based authentication, Adatum also enables several future capabilities for user authentication. For example, Adatum may decide in future to authenticate users for aExpense or other applications by accepting claims issued by a social identity provider such as a Microsoft account or Google. By using the Windows Azure Access Control service, and connecting it to their own on-premises ADFS, Adatum can provide a wide-ranging federated authentication mechanism that supports single-sign-on (SSO) for users through many common identity providers as well as through Adatum's own Active Directory.

This will remove the direct dependency that the current version of the application has on Active Directory because the application will obtain the required user data from the claims issuer (the claims issuer still has to get the data from Active Directory on behalf of the aExpense application). The external claims issuer can integrate with Active Directory, so that application users will continue to have the same single sign-on experience.

Using claims can simplify the application by delegating responsibilities to the claims issuer.

Changing to use claims-based authentication will mean that the developers must modify the application. However, as they will need to refactor it as a Cloud Services solution, the additional work required was considered to be acceptable in view of the ability to remove the reliance on a direct connection back to their on-premises Active Directory.

The example application provided with this guide uses a mock issuer so that you can run the example on a single workstation without needing to set up Active Directory and ADFS. The mock issuer is also used by Adatum when testing the application. For more details see the section "Using a Mock Issuer" later in this chapter.

Profile Management

The on-premises aExpense application stores users' preferred reimbursement methods by using the ASP.NET profiles feature. When migrating the application to Windows Azure using virtual machines, Adatum chose to keep this mechanism by creating a suitable ASPNET-DB database in the SQL Server hosted in the cloud. This minimized the changes required to the application at that stage.

At a later stage of the migration, the team will use a profile provider implementation that uses Windows Azure table storage. For more details, see Chapter 7, "Moving to Windows Azure Table Storage," of this guide.

Session Data Management

The AddExpense.aspx page uses session state to maintain a list of expense items before the user saves the completed business expense submission.

The on-premises aExpense application stores session data in-memory using the standard ASP.NET session mechanism. This works well when there is only a single instance of the application because every request from a user will be handled by this single server. However, if Adatum decides to take advantage of the elasticity feature of Windows Azure to deploy additional instances of the application (a major goal for the migration to Windows Azure), the developers must consider how this will affect session data storage. If Adatum uses more than a single instance of the web application, the session state storage mechanism must be web farm friendly, so that the session state data is accessible from every instance.

There are several options for managing shared session state in Windows Azure.

If we were building a new application that required profile storage, we would consider using the Universal Providers, which are the recommended providers in Visual Studio for many types of data access. The Universal Providers can also be used to store session state.

Storing Session State Data in a Database

The *Microsoft ASP.NET Universal Providers* enable you to store your session state in either SQL Server or SQL Database. At the time of writing, these providers are available through *NuGet*. After you have installed the package, you can use the providers by modifying your configuration.

To use this option you must have a SQL Database subscription or have SQL Server installed on a Windows Azure virtual machine. Therefore, it is most cost effective if you are already using either SQL Server or SQL Database.

Adatum does have a SQL Server database available that could be used to store session state, but considered that in future it may wish to explore how it can move away from using a relational database altogether. For this reason, Adatum chose not to adopt this option for storing session state because it would add an additional dependency on SQL Server.

Storing Session State Data in Windows Azure Storage

The *Windows Azure ASP.NET Providers* enable you to store membership, profile, role, and session state information in Windows Azure table and blob storage. These providers are published as sample code on the Windows Azure developer samples site.

Adatum will have a suitable storage account available for these providers to use. However, Adatum is concerned that the providers are samples, and are still under development. In addition, Adatum is concerned that stored session state may not be removed automatically from blobs when sessions are abandoned. Therefore Adatum chose not to use these providers for storing session state.

Storing Session State Data in a Windows Azure Cache

The third option is to use the *ASP.NET 4 Caching Providers for Windows Azure*. This option enables you to use Windows Azure Caching to store your application session state. In most scenarios, this option will provide the best performance, and at the time of writing this is the only officially supported option.

The ASP.NET 4 Caching Providers for Windows Azure work with two different types of cache in Windows Azure:

- **Windows Azure Caching**. This is a high-performance, distributed, in-memory cache that uses memory from your Cloud Services roles. You configure this cache as part of the deployment of your application so that this cache is private to the deployment. You can specify whether to use dedicated caching roles, where role instances simply provide memory and resources for the cache, or you can allocate a proportion of each role instance that also run application code for caching. There is no separate charge for using this type of cache; you pay the standard rates for running the Cloud Services role instances that host the cache.

- **Windows Azure Shared Caching**. This is a high-performance, distributed, in-memory caching service that you provision separately from your other Windows Azure services, and that all of the services in your subscription can use. You pay for this shared caching on monthly based on the size of the cache (for example, at the time of writing, a 128 MB cache costs $ 45.00 per month). For current pricing information, see the Windows Azure *Pricing Details*.

    For more information about Caching in Windows Azure and the differences between Windows Azure Caching and Windows Azure Shared Caching see, "Caching in Windows Azure" on MSDN.

Adatum decided that the ASP.NET 4 Caching Providers for Windows Azure provide the ideal solution for the aExpense application because they are easy to integrate into the application, they are supported in the compute emulator so that development and testing are simpler (developers can run and debug the application entirely within the emulator instead of setting up the cache in the cloud), and they will not hold extraneous session state data after the session expires.

Adatum will modify the aExpense configuration to use these providers for storing session state, and use a co-located cache in the web role instances of the application in order to minimize the running costs for the application. However, Adatum must ensure that it configures a sufficiently large cache so that session data is not evicted as the cache fills up.

As an alternative to using session state to store the list of expense items before the user submits them, Adatum considered using ASP.NET view state so that the application maintains its state data on the client. This solution would work when the application has multiple web role instances because the application does not store any state data on the server. Because later versions of the aExpense application store scanned images in the state before the application saves the whole expense submission, this means that the state can be quite large. Using view state would be a poor solution in this case because it would need to move the data in the view state over the network, using up bandwidth and adversely affecting the application's performance.

Using ASP.NET view state is an excellent solution as long as the amount of data involved is small. This is not the case with the aExpense application where the state data will include images.

Data Storage

Adatum also reviewed the choice made during the initial migration to use a hosted SQL Server running in a Windows Azure virtual machine as the data store for the aExpense application. There is no technical requirement to change this at the moment because code running in Cloud Services roles can connect to the database in exactly the same way as in the virtual machine hosted version of the application.

This does not mean that Adatum will not reconsider the choice of database deployment in the future, but it is not mandatory for this step and so the migration risks are reduced by delaying this decision to a later phase.

> *In Chapter 4, "Moving to Windows Azure SQL Database," of this guide you will discover how Adatum revisited the decision to use a hosted SQL Server, and chose to move to the PaaS equivalent - Windows Azure SQL Database.*

However, the IaaS solution used Windows Authentication to connect to SQL Server and Adatum plans to remove the requirement for a virtual network, so the application will now need to use SQL authentication. This will require configuration changes in both the application and SQL Server.

Now that the connection string in the configuration file includes credentials in plain text, you should consider encrypting this section of the file. This will add to the complexity of your application, but it will enhance the security of your data. If your application is likely to run on multiple role instances, you must use an encryption mechanism that uses keys shared by all the role instances.

> *To encrypt your SQL connection string in the Web.config file, you can use the **Pkcs12 Protected Configuration Provider**.*

> *For additional background information about using this provider, see the sections "Best Practices on Writing Secure Connection Strings for SQL Database" and "Create and deploy an ASP.NET application with Security in mind" in the post **"Windows Azure SQL Database Connection Security."***

Application Configuration

The on-premises version of the aExpense application uses the Web.config file to store configuration settings, such as connection strings and authentication information for the application. When an application is deployed to Windows Azure Cloud Services it's not easy to edit the Web.config file. You must redeploy the application when values need to be changed. However, it is possible to edit the service configuration file through the portal or by using a PowerShell script to make configuration changes on the fly. Therefore, Adatum would like to move some configuration settings from the Web.config file to the service configuration file (ServiceConfiguration.csfg).

However, some components of an application may not be "cloud aware" in terms of reading configuration settings from the Windows Azure service configuration files. For example, the ASP.NET Profile Provider that Adatum uses to store each user's preferred reimbursement method will only read the connection string for its database from the Web.config file. The Windows Identity Foundation (WIF) authentication mechanism also depends on settings that are located in the Web.config file.

> When using Windows Azure, the recommendation is to store all application settings in the service configuration file. You can edit this file at any time without redeploying the application, and changes do not force an application restart. However, at the time of writing, the ASP.NET providers such as the membership, profile, role, and the Universal Providers, will only recognize a database connection string stored in the Web.config file.

To resolve this, the developers at Adatum considered implementing code that runs when the application starts to copy values from the ServiceConfiguration.csfg file to the active configuration loaded from the Web.config file. To assess whether this was a viable option, the developers needed to explore how startup tasks can be executed in a Windows Azure Cloud Services role.

Application Startup Processes

The developers at Adatum considered that they may want to execute some code that runs only as the application starts. This prompted them to investigate how to execute tasks when a Windows Azure application starts up. The processes that occur when a Windows Azure web or worker role is started are:

1. The Windows Azure fabric controller creates and provisions the virtual machine that will run the application and loads the application code.

2. The fabric controller looks for startup tasks defined in the ServiceConfiguration.cscfg file, and starts any it finds in the order they are defined in the file.

3. The fabric controller fires the **OnStart** event in the **RoleEntry-Point** class when the role starts executing.

4. The fabric controller fires the **Run** event in the **RoleEntryPoint** class when the role is ready to accept requests.

5. Global application events such as **Application_Start** in Global.asax are fired in ASP.NET web roles.

Startup tasks are command line executable programs that can be executed either asynchronously or synchronously. Asynchronous execution can be **Foreground** (the role cannot shut down until the task completes) or **Background** ("fire and forget"). Startup tasks can also be executed with a **Limited** permission level (the same as the role instance) or **Elevated** (with administrative permissions). Startup tasks are typically used for executing code that accesses features outside of the role, installing and configuring software and services, and for tasks that require administrative permission.

The two events that can be handled in the RoleEntryPoint class of web or worker role occur when the role has completed loading and is about to start (**OnStart)**, and when startup is complete and the role is ready to accept requests (**Run**). The **OnStart** and **Run** events are typically used to execute .NET code that prepares the application for use; for example, by loading data or preparing storage.

If the code needs additional permissions, the role can be started in **Elevated** mode. The **OnStart** method will then run with administrative permissions, but the remainder of the role execution will revert to the more limited permission level.

Use startup tasks to run executable programs when your role starts, and use the **OnStart** and **Run** events to execute .NET code within a role after it loads and when it is ready to accept requests. In a web role, you can also use the **Application_Start** method in the Global.asax file

You must be careful not to access the request context or call methods of the **RoleManager** *class from the* **Application_ Start** *method. For further details, see the RoleManager class documentation on MSDN.*

Keep in mind that a role may be restarted after a failure or during an update process, and so the **OnStart** and **Run** events may occur more than once for a role instance. Also remember that asynchronous startup tasks may still be executing when the **OnStart** and **Run** events occur, which makes it important to ensure that they cannot cause a race or deadlock condition. For more information about using startup tasks see *"Real World: Startup Lifecycle of a Windows Azure Role"* and *"Running Startup Tasks in Windows Azure."*

Copying Configuration Values in a Startup Task

There are several workarounds you might consider implementing if you decide to copy values from the service configuration file to the active Web.config configuration. Adatum's developers carried out tests to evaluate their capabilities.

One approach is to run code in the **OnStart** event of a web role when the role is started with elevated permissions. The post *"Edit and Apply New WIF's Config Settings in Your Windows Azure Web Role Without Redeploying"* describes this approach.

Another is to execute code in a startup task that uses the AppCmd.exe utility to modify the configuration before the role starts. The page *"How to: Use AppCmd.exe to Configure IIS at Startup"* describes how this approach can be used to set values in configuration.

There is also an issue in the current release of Windows Azure regarding the availability of the Identity Model assembly. See *"Unable to Find Assembly 'Microsoft.IdentityModel' When RoleEnvironmentAPIs are Called"* for more information.

After considering the options and the drawbacks, the developers at Adatum decided to postpone implementing any of the workarounds in the current version of the aExpense application. It is likely that future releases of Windows Azure will resolve the issues and provide a recommended approach for handling configuration values that would normally reside in the Web.config file.

Windows Azure roles also support the **OnStop** event that fires when a role is about to stop running. You can handle this event to perform cleanup tasks for the role such as flushing queues, releasing resources, forcing completion of running tasks, or executing other processes. Your handler code, and all the processes it executes, must complete within five minutes.

Solution Summary

Figure 1 shows the whiteboard drawing that the team used to explain the architecture of aExpense after this step of the migration to Windows Azure.

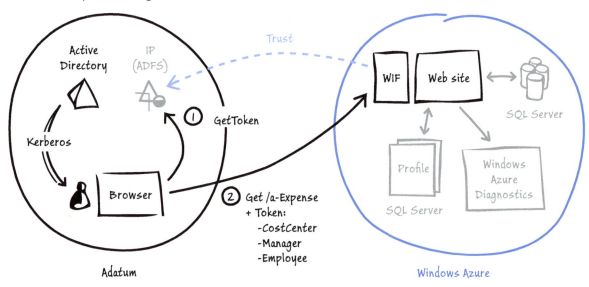

FIGURE 1
aExpense as an application hosted in Windows Azure

> Many of the changes you must make to the application before deploying it to Windows Azure apply to both deploying as a Windows Azure Web Site and deploying to a Cloud Service. For example, you must ensure that session state can be maintained correctly across all instances, and that profile information and authentication mechanisms are compatible with the Windows Azure hosted services approach.

INSIDE THE IMPLEMENTATION

Now is a good time to walk through the process of migrating aExpense into a cloud-based application in more detail. As you go through this section, you may want to download the Visual Studio solution from *http://wag.codeplex.com/.* This solution contains an implementation of the aExpense application (in the **Azure-CloudService-SQLServer** folder) after the migration step described in this chapter. If you are not interested in the mechanics, you should skip to the next section.

The Hands-on Labs that accompany this guide provide a step-by-step walkthrough of parts of the implementation tasks Adatum carried out on the aExpense application at this stage of the migration process.

Creating a Web Role

The developers at Adatum created the Visual Studio solution for the cloud-based version of aExpense by using the Windows Azure Cloud Service template. This template generates the required service configuration and service definition files, and the files for the web and worker roles that the application will need. For more information on this process, see "*Creating a Windows Azure Project with Visual Studio.*"

This first cloud-based version of aExpense has a single web role that contains all the code from the original on-premises version of the application.

The service definition file defines the endpoint for the web role. The aExpense application only has a single HTTPS endpoint, which requires a certificate. In this case, it is known as "localhost." When you deploy the application to Windows Azure, you'll also have to upload the certificate. For more information, see "*Configuring SSL for an Application in Windows Azure.*"

Use the Visual Studio Windows Azure Cloud Service template from the Cloud section in the New Project dialog to get started with your cloud project.

```xml
XML
<ServiceDefinition name="aExpense.Azure" xmlns="…">
  <WebRole name="aExpense" vmsize="Medium">
    <Sites>
      <Site name="Web">
        <Bindings>
          <Binding name="Endpoint1" endpointName="Endpoint1" />
        </Bindings>
      </Site>
    </Sites>
    <Endpoints>
      <InputEndpoint name="Endpoint1" protocol="https" port="443"
        certificate="localhost" />
    </Endpoints>
    <Certificates>
      <Certificate name="localhost" storeLocation="LocalMachine"
        storeName="My" />
    </Certificates>
    <ConfigurationSettings>
      <Setting name="DataConnectionString" />
    </ConfigurationSettings>
    <Imports>
      <Import moduleName="Diagnostics" />
    </Imports>
    <LocalResources>
      <LocalStorage name="DiagnosticStore"
        cleanOnRoleRecycle="false" sizeInMB="20000" />
    </LocalResources>
  </WebRole>
</ServiceDefinition>
```

The "localhost" certificate is only used for testing your application.

The service configuration file defines the aExpense web role. It contains the connection strings that the role will use to access storage and details of the certificates used by the application. The application uses the **DataConnectionString** to connect to the Windows Azure storage holding the profile data, and uses the **DiagnosticsConnectionString** to connect to the Windows Azure storage for saving logging and performance data. The connection strings will need to change when you deploy the application to the cloud so that the application can use Windows Azure storage.

```xml
XML
<ServiceConfiguration serviceName="aExpense.Azure" xmlns="…">
  <Role name="aExpense">
    <Instances count="1" />
    <ConfigurationSettings>
      <Setting name=
    "Microsoft.WindowsAzure.Plugins.Diagnostics.ConnectionString"
              value="DefaultEndpointsProtocol=https;
              AccountName={Azure storage account name};
              AccountKey={Azure storage shared key}" />
      <Setting name="DataConnectionString"
              value="DefaultEndpointsProtocol=https;
              AccountName={Azure storage account name};
              AccountKey={Azure storage shared key}" />
    </ConfigurationSettings>
    <Certificates>
      <Certificate name="localhost" thumbprint="…"
                thumbprintAlgorithm="sha1" />
    </Certificates>
  </Role>
</ServiceConfiguration>
```

The values of "Azure storage account name" and "Azure storage shared key" are specific to your Windows Azure storage account.

Reading Configuration Information

In the original on-premises application, settings such as connection strings are stored in the Web. config file. Configuration settings for Windows Azure Cloud Services (web roles and worker roles) are stored in the ServiceConfiguration.cscfg and ServiceDefinition.csdef files. This allows, amongst other benefits, easy modification of the configuration settings by using the Windows Azure Portal or Power-Shell cmdlets; without needing to redeploy the entire application.

To facilitate testing the aExpense application when it runs in the local emulator, the developers at Adatum created the aExpense.Azure project with two service configuration files; one contains a connection string for a local SQL Server instance, and one contains the connection string for the test SQL Server instance hosted in the cloud. This makes it easy to switch between these configurations in Visual Studio without the need to edit the configuration files whenever the deployment target changes.

To find out more about using multiple service configuration files in a Windows Azure project, see "How to: Manage Multiple Service Configurations for a Windows Azure Application."

The developers at Adatum wanted to include all of the application configuration settings, such as database connection strings and other values, in the service configuration files and not in the Web.config file. However, due to the issues that arise when attempting to achieve this (see the section "Application Configuration" earlier in this chapter) they decided not to implement it in the current version of the aExpense application. They will revisit this decision as new releases of the Windows Azure Cloud Services platform become available.

This means that the application must read values from both the service configuration file and the Web.config file. By using the Windows Azure **CloudConfigurationManager** class to read configuration settings, the application will automatically look first in the ServiceConfiguration.cscfg file. This means that the application code will read connection strings that specify the location of the Expenses data, and other settings, from the ServiceConfiguration.cscfg file.

Using the Windows Azure CloudConfigurationManager Class

The **CloudConfigurationManager** class simplifies the process of reading configuration settings in a Windows Azure application because the methods automatically read settings from the appropriate location. If the application is running as a .NET web application they return the setting value from the Web.config or App.config file. If the application is running as Windows Azure Cloud Service or as a Windows Azure Web Site, the methods return the setting value from the ServiceConfiguration.cscfg or ServiceDefinition.csdef file. If the specified setting is not found in the service configuration file, the methods look for it in the Web.config or App.config file.

However, the "fall through" process the **CloudConfigurationManager** class uses when it cannot find a setting in the ServiceConfiguration.cscfg or ServiceDefinition.csdef file only looks in the **<appSettings>** section of the Web.config or App.config file. Connection strings, and many other settings, are not located in the **<appSettings>** section. To resolve this, Adatum created a custom class named **CloudConfiguration** that uses the **CloudConfigurationManager** class internally. For example, the following code shows the **GetConnectionString** method in the custom **CloudConfiguration** class.

```csharp
C#
public static string GetConnectionString(string settingName)
{
  // Get connection string from the service configuration file.
  var connString
      = CloudConfigurationManager.GetSetting(settingName);
  if (string.IsNullOrWhiteSpace(connString))
  {
    // Fall back to connectionStrings section in Web.config.
    return ConfigurationManager.ConnectionStrings[
                          settingName].ConnectionString;
  }
  return connString;
}
```

> The **CloudConfiguration** class is located in the **Shared\aExpense** folder of the examples available for this guide.

Implementing Claims-based Authentication

Before this step of Adatum's migration process, aExpense used Windows Authentication to authenticate users. This is configured in the Web.config file of the application. In this step, the aExpense application delegates the process of validating credentials to an external claims issuer instead of using Windows Authentication. You make this configuration change in the Web.config file.

> *To find out more about claims-based Identity, the **FedUtil** tool, and Windows Identity Foundation (WIF) take a look at the book "A Guide to Claims-Based Identity and Access Control". You can download a .pdf copy of this book.*

The first thing that you'll notice in the Web.config file is that the authentication mode is set to **None**, while the requirement for all users to be authenticated has been left in place.

XML
```
<authorization>
  <deny users="?" />
</authorization>
<authentication mode="None" />
```

The **WSFederationAutheticationModule** (FAM) and **SessionAuthenticationModule** (SAM) modules now handle the authentication process. You can see how these modules are loaded in the **system. webServer** section of the Web.config file.

You can make these changes to the Web.config file by running the FedUtil tool.

XML
```
<system.webServer>
  …
  <add name="WSFederationAuthenticationModule"
      type="Microsoft.IdentityModel.Web.
          WSFederationAuthenticationModule, …" />
  <add name="SessionAuthenticationModule"
      type="Microsoft.IdentityModel.Web.
          SessionAuthenticationModule, …" />
</system.webServer>
```

When the modules are loaded, they're inserted into the ASP.NET processing pipeline in order to redirect the unauthenticated requests to the claims issuer, handle the reply posted by the claims issuer, and transform the security token sent by the claims issuer into a **Claims-Principal** object. The modules also set the value of the **HttpContext. User** property to the **ClaimsPrincipal** object so that the application has access to it.

More specifically, the **WSFederationAuthenticationModule** redirects the user to the issuer's logon page. It also parses and validates the security token that is posted back. This module also writes an encrypted cookie to avoid repeating the logon process. The **SessionAuthenticationModule** detects the logon cookie, decrypts it, and repopulates the **ClaimsPrincipal** object. After the claim issuer authenticates the user, the aExpense application can access the authenticated user's name.

The Web.config file contains a new section for the **Microsoft.IdentityModel** that initializes the Windows Identity Foundation (WIF) environment.

```XML
<microsoft.identityModel>
  <service>
    …
  </service>
</microsoft.identityModel>
```

You can also use a standard control to handle the user logout process from the application. The following code example from the Site.Master file shows a part of the definition of the standard page header.

```HTML
<div id="toolbar">
    Logged in as:
    <i>
      <%= Microsoft.Security.Application.Encoder.HtmlEncode
          (this.Context.User.Identity.Name) %>
    </i> |
    <idfx:FederatedPassiveSignInStatus
          ID="FederatedPassiveSignInStatus1"
          runat="server"
          OnSignedOut="FederatedPassiveSignInStatus1SignedOut"
          SignOutText="Logout" FederatedPassiveSignOut="true"
          SignOutAction="FederatedPassiveSignOut" />
</div>
```

You'll also notice a small change in the way that aExpense handles authorization. Because the authentication mode is now set to **None** in the Web.config file, the authorization rules in the Web.config file now explicitly deny access to all users as well as allowing access for the designated role.

```xml
XML
<location path="Approve.aspx">
  <system.web>
    <authorization>
      <allow roles="Manager" />
      <deny users="*"/>
    </authorization>
  </system.web>
</location>
```

The claim issuer now replaces the ASP.NET role management feature as the provider of role membership information to the application.

There is one further change to the application that potentially affects the authentication process. If you were to run the aExpense application on more than one web role instance in Windows Azure, the default cookie encryption mechanism (which uses DPAPI) is not appropriate because each instance has a different key. This would mean that a cookie created by one web role instance would not be readable by another web role instance. To solve this problem you should use a cookie encryption mechanism that uses a key shared by all the web role instances. The following code from the Global.asax file shows how to replace the default **SessionSecurityHandler** object and configure it to use the **RsaEncryptionCookieTransform** class.

Although the initial deployment of aExpense to Windows Azure will only use a single web role, we need to make sure that it will continue to work correctly when we scale up the application. That is why we use RSA with a certificate to encrypt the session cookie.

```csharp
C#
private void OnServiceConfigurationCreated(object sender,
    ServiceConfigurationCreatedEventArgs e)
{
    // Use the <serviceCertificate> to protect the cookies that
    // are sent to the client.
    List<CookieTransform> sessionTransforms =
        new List<CookieTransform>(
            new CookieTransform[]
            {
                new DeflateCookieTransform(),
                new RsaEncryptionCookieTransform(
                    e.ServiceConfiguration.ServiceCertificate),
                new RsaSignatureCookieTransform(
                    e.ServiceConfiguration.ServiceCertificate)
            });
    SessionSecurityTokenHandler sessionHandler =
    new
     SessionSecurityTokenHandler(sessionTransforms.AsReadOnly());

    e.ServiceConfiguration.SecurityTokenHandlers.AddOrReplace(
        sessionHandler);
}
```

Managing User Data

Before the migration, aExpense used an LDAP query to retrieve Cost Center, Manager, and Display Name information from Active Directory. It used the ASP.NET Role provider to retrieve the role membership of the user, and the ASP.NET Profile Provider to retrieve the application specific data for the application—in this case, the preferred reimbursement method. The following table summarizes how aExpense accesses user data, and where the data is stored before the migration:

User Data	Access Mechanism	Storage
Role Membership	ASP.NET Role Provider	SQL Server
Cost Center	LDAP	Active Directory
Manager	LDAP	Active Directory
Display Name	LDAP	Active Directory
User Name	ASP.NET Membership Provider	SQL Server
Preferred Reimbursement Method	ASP.NET Profile Provider	SQL Server

After the migration, aExpense continues to use the same user data, but it accesses the data differently. The following table summarizes how aExpense accesses user data, and where the data is stored after the migration:

User Data	Access Mechanism	Storage
Role Membership	ADFS	Active Directory
Cost Center	ADFS	Active Directory
Manager	ADFS	Active Directory
Display Name	ADFS	Active Directory
User Name	ADFS	Active Directory
Preferred Reimbursement Method	ASP.NET Profile Provider	SQL Server

The external issuer delivers the claim data to the aExpense application after it authenticates the application user.

The external issuer delivers the claim data to the aExpense application after it authenticates the application user. The aExpense application uses the claim data for the duration of the session and does not need to store it.

The application can read the values of individual claims whenever it needs to access claim data. You can see how to do this if you look in the **ClaimHelper** class.

Managing Session Data

To switch from using the default, in-memory session provider to the web farm friendly ASP.NET 4 Caching Providers for Windows Azure, the developers made two configuration changes.

The first change is in the ServiceConfiguration.cscfg file to specify the size and the location of the class. The following snippet shows how Adatum allocates 15% of the memory of each role instance to use as a distributed cache. The value for **NamedCaches** is the default set by the SDK; it allows you to change the cache settings while the application is running simply by editing the configuration file.

```XML
<ConfigurationSettings>
  ...
  <Setting
    name="Microsoft.WindowsAzure.Plugins.Caching.NamedCaches"
    value="{&quot;caches&quot;:[{&quot;name&quot;:&quot;default
           &quot;,&quot;policy&quot;:{&quot;eviction&quot;
           :{&quot;type&quot;:0},&quot;expiration&quot;
           :{&quot;defaultTTL&quot;:10,&quot;isExpirable&quot;
           :true,&quot;type&quot;:1},&quot;
            serverNotification&quot;:{&quot;isEnabled&quot;
           :false}},&quot;secondaries&quot;:0}]}" />
  <Setting
    name="Microsoft.WindowsAzure.Plugins.Caching.DiagnosticLevel"
    value="1" />
  <Setting
    name="Microsoft.WindowsAzure.Plugins.Caching.Loglevel"
    value="" />
  <Setting name=
    "Microsoft.WindowsAzure.Plugins.Caching.CacheSizePercentage"
    value="15" />
  <Setting name=
    "Microsoft.WindowsAzure.Plugins.Caching
    .ConfigStoreConnectionString"
    value="UseDevelopmentStorage=true" />
</ConfigurationSettings>
```

Remember that you can use the multiple service configuration feature in Visual Studio to maintain multiple versions of your service configuration file. For example, you can have one configuration for local testing and one configuration for testing your application in Windows Azure.

In Visual Studio, you can use the property sheet for the role to set these values using a GUI rather than editing the values directly. When you deploy to Windows Azure you should replace the use of development storage with your Windows Azure storage account. For more information, see *Configuration Model* on MSDN.

The second configuration change is in the Web.config file to enable the ASP.NET 4 Caching Providers for Windows Azure. The following snippet shows how Adatum configured the session provider for the aExpense application.

```XML
<sessionState mode="Custom" customProvider="NamedCacheBProvider">
  <providers>
    <add cacheName="default" name="NamedCacheBProvider"
      dataCacheClientName="default" applicationName="aExpense"
      type="Microsoft.Web.DistributedCache
      .DistributedCacheSessionStateStoreProvider,
      Microsoft.Web.DistributedCache" />
  </providers>
</sessionState>
```

For more information, see *ASP.NET Session State Provider Configuration Settings* on MSDN.

TESTING, DEPLOYMENT, MANAGEMENT, AND MONITORING

This section contains topics that describe the way that Adatum needed to review its testing, deployment, management, and monitoring techniques when the aExpense application was deployed to Windows Azure using the PaaS approach and Cloud Services.

The techniques that Adatum needed to review are how pre-deployment testing is carried out, how the application is packaged and deployed, and the ways that Adatum's administrators can manage and monitor the application when it is hosted in Windows Azure Cloud Services.

Testing Cloud Services Applications

When you're developing a Cloud Services application for Windows Azure, it's best to do as much development and testing as possible by using the local compute emulator and storage emulator. At Adatum, developers run unit tests and ad-hoc tests in the compute emulator and storage emulator running on their local computers. Although the emulators are not identical to the cloud environment, they are suitable for developers to run tests on their own code. The build server also runs a suite of tests as a part of the standard build process. This is no different from the normal development practices for on-premises applications.

Moving to using Windows Azure Cloud Services meant that we needed to review our existing processes for testing, deploying, managing, and monitoring applications.

Most testing can be performed using the compute emulator and storage emulator.

The testing team performs the majority of its tests using the local compute emulator as well. They only deploy the application to the Windows Azure test environment to check the final version of the application before it is passed to the administration team for deployment to production. This way, they can minimize the costs associated with the test environment by limiting the time that they have a test application deployed in the cloud.

However, there are some differences between the compute and storage emulators and the Windows Azure runtime environment, and so local testing can only provide the platform for the first level of testing. The final pre-production testing must be done in a real Windows Azure environment. This is accomplished by using the separate test and production environments in the cloud.

> *For details of the differences between the local emulator environment and Windows Azure, see* **Differences Between the Compute Emulator and Windows Azure** *and* **Differences Between the Storage Emulator and Windows Azure Storage Services** *on MSDN.*

You can deploy an application to your Windows Azure test environment just while you run the tests, but don't forget that any time you do something in Windows Azure–even if it's only testing–it costs money. You should remove test instances when you are not using them.

Cloud Services Staging and Production Areas

Adatum wants to be able to deploy an application to either a staging or a production area. Windows Azure Cloud Services provides both a staging and a production area for roles you deploy; you can deploy an application to either a staging or a production environment within the same Cloud Service. A common scenario is to deploy first to the staging environment and then, at the appropriate time, move the new version to the production environment. The only difference is in the URL you use to access them.

In the staging environment the URL to access the aExpense web role will be something obscure like http://aExpenseTestsjy6920asgd09. cloudapp.net, while in the production environment you will have a friendly URL such as http://aExpense.cloudapp.net. This allows you to test new and updated applications in a private environment that others don't know about before deploying them to production. You can also swap the contents of the production and staging areas, which makes it easy to deploy or roll back the application to the previous version without needing to redeploy the role.

You can quickly deploy or roll back a change in production by using the swap operation. The swap is almost instantaneous because it just involves Windows Azure changing the DNS entries for the two areas.

This feature is useful for Adatum to perform no downtime upgrades in the production environment. The operations staff can deploy a new version of the aExpense application to the staging deployment slot, perform some final tests, and then swap the production and staging slots to make the new version of the application available to users.

All Windows Azure environments in the cloud are the same; there's nothing special about the staging area or a separate test subscription. You can be sure that different subscriptions have identical environments—this is something that's very difficult to guarantee in your on-premises environments.

Because development and testing staff don't have access to the production environment, there's no risk of accidentally deploying to the live environment.

Having separate Windows Azure subscriptions for production and test helps Adatum to track the costs of running the application in production separately from the costs associated with development and test.

However, Adatum wants to control access to the live production environment so that only administrators can deploy applications there. To achieve this, Adatum uses separate Windows Azure subscriptions.

Separate Test and Live Subscriptions

To ensure that only administrators have access to the live production environment, Adatum has two Windows Azure subscriptions. One is used for testing, and one is the live production subscription. The two subscriptions are standard Windows Azure subscriptions, and so provide identical environments. Adatum can be confident that the application will run in the same way in both of them. Adatum can also be sure that deployment of the application will work in the same way in both environments because it can use the same package to deploy to both test and production.

The test application connects to a test database server running in the test subscription so that developers and testers can manage the data used for testing. The live application running in the production environment uses a different server name in the connection string to connect to the live database that also runs in the production environment.

Because each subscription has its own Microsoft account and its own set of API keys, Adatum can limit access to each environment to a particular set of personnel. Members of the testing team and key members of the development team have access to only the testing account. Only two key people in the Adatum operations department have access to the production account.

Microsoft bills Adatum separately for the two environments, which makes it easy for Adatum to separate the running costs for the live environment from the costs associated with development and test. This allows Adatum to manage the product development budget separately from the operational budget, in a manner similar to the way it manages budgets for on-premises applications.

Figure 2 summarizes the application life cycle management approach at Adatum.

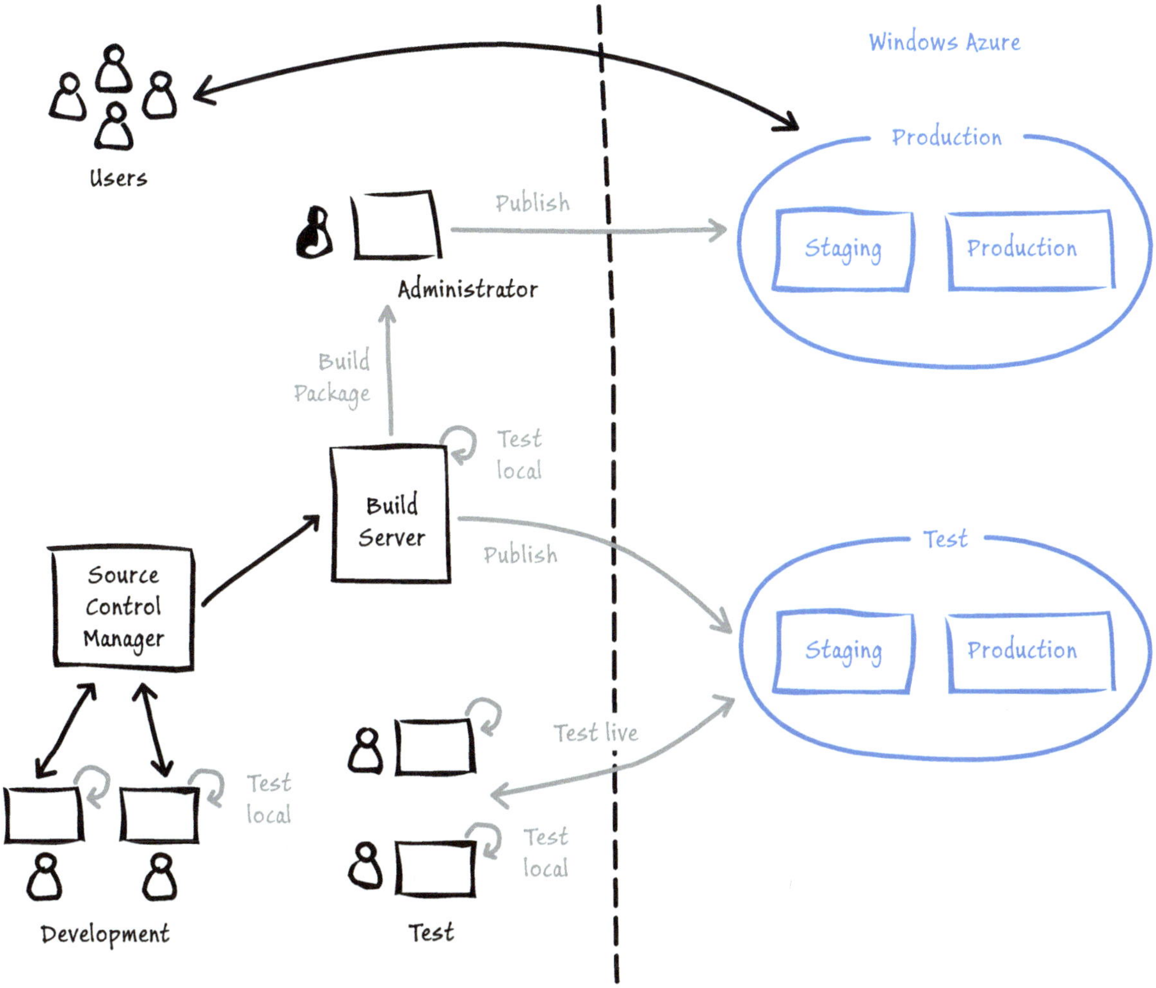

FIGURE 2
Adatum's application life cycle management environment

Figure 2 shows the two separate Windows Azure subscriptions that Adatum uses for test and production, as well as the on-premises environment that consists of development computers, test computers, a build server, and a source code management tool.

We use scripts because it means that we can reliably repeat operations. We're also planning to enhance some of our scripts to record when they were run, by whom, and what parameters were used. See the section "Using Deployment Scripts" later in this chapter for more details.

After you have deployed your application to Windows Azure, it is possible to edit the settings in the service configuration file while the application is running to make changes on the fly. You should think carefully about what configuration settings you put in the .cscfg file for your application.

Managing Windows Azure Services

There are three ways to access a Windows Azure environment to perform management tasks such as deploying and removing roles and managing other services. The first is through the Windows Azure Management Portal, where a single Microsoft account has access to everything in the portal. The second is by using the Windows Azure Service Management API, where API certificates are used to access to all the functionality exposed by the API. The third is to use the Windows Azure Management PowerShell cmdlets. In all three cases there is currently no way to restrict users to only be able to manage a subset of the available functionality. Within Adatum, almost all operations that affect the test or production environment are performed using scripts based on the PowerShell cmdlets instead of the management portal.

Setup and Deployment

This section describes how Adatum configures, packages, and deploys the aExpense application. To deploy an application to Windows Azure Cloud Services you must upload two files: the service package file that contains all your application's files, and the service configuration file (.cscfg) that contains your application's configuration data. You can generate the service package file either by using the Cspack. exe command-line utility or by using Visual Studio if you have installed the Windows Azure SDK.

Managing Different Local, Test, and Live Configurations

Adatum uses different configuration files when deploying to the local Compute Emulator, test, and live (production) environments. For the aExpense application, the key difference between the contents of these configuration files is the storage connection strings. In Windows Azure storage, this information is unique to each Windows Azure subscription and uses randomly generated access keys; for SQL Server or SQL Database the connection string includes the user name and password to connect to the database.

The developers and testers at Adatum make use of the feature in Visual Studio that allows you to maintain multiple service configuration files. Adatum includes one service configuration for testing on the local Compute and Storage Emulators, and another configuration for deploying to the test subscription in Windows Azure, as shown in Figure 3.

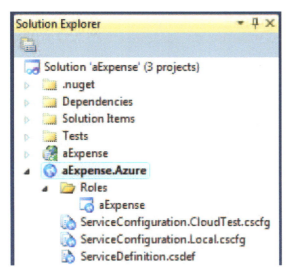

FIGURE 3
Using separate configuration files for local and cloud configurations

This is what the relevant section of the "Local" service configuration file looks like. It is used when the application is running in the local Compute Emulator and using the local Storage Emulator.

```xml
<ConfigurationSettings>
  <Setting name="DataConnectionString"
    value="UseDevelopmentStorage=true" />
  <Setting name="Microsoft.WindowsAzure.Plugins
                .Diagnostics.ConnectionString"
    value="UseDevelopmentStorage=true" />
  <Setting name="Microsoft.WindowsAzure.Plugins
                .Caching.NamedCaches"
    value="" />
  <Setting name="Microsoft.WindowsAzure.Plugins
                .Caching.Loglevel"
    value="" />
  <Setting name="Microsoft.WindowsAzure.Plugins
                .Caching.CacheSizePercentage"
    value="15" />
  <Setting name="Microsoft.WindowsAzure.Plugins
                .Caching.ConfigStoreConnectionString"
    value="UseDevelopmentStorage=true" />
  <Setting name="aExpense"
    value="Data Source=LocalTestSQLServer;
           Initial Catalog=aExpense;
           Integrated Security=True" />
</ConfigurationSettings>
```

This is what the same section of the "CloudTest" service configuration file looks like. It specifies the resources within Adatum's Windows Azure test subscription that the application will use when deployed there.

```XML
<ConfigurationSettings>
  <Setting name="DataConnectionString"
    value="DefaultEndpointsProtocol=https;
          AccountName={StorageAccountName};
          AccountKey={StorageAccountKey}" />
  <Setting name="Microsoft.WindowsAzure.Plugins
              .Diagnostics.ConnectionString"
    value="DefaultEndpointsProtocol=https;
          AccountName={StorageAccountName};
          AccountKey={StorageAccountKey}" />
  <Setting name="Microsoft.WindowsAzure.Plugins
              .Caching.NamedCaches"
    value="" />
  <Setting name="Microsoft.WindowsAzure.Plugins
              .Caching.Loglevel"
    value="" />
  <Setting name="Microsoft.WindowsAzure.Plugins
              .Caching.CacheSizePercentage"
    value="15" />
  <Setting name="Microsoft.WindowsAzure.Plugins
              .Caching.ConfigStoreConnectionString"
    value="DefaultEndpointsProtocol=https;
          AccountName={StorageAccountName};
          AccountKey={StorageAccountKey}" />
  <Setting name="aExpense"
    value="Data Source=CloudTestSQLServer;
          Initial Catalog=aExpenseTest;
          UId={UserName};Pwd={Password};" />
</ConfigurationSettings>
```

The values of the storage account name, storage account key, and database login credentials are specific to the Windows Azure storage account and the cloud-hosted database. The configuration values in the examples available for this guide differ from those shown above. If you want to edit these settings, you should use the tools in Visual Studio instead of directly editing the XML.

Notice that the project does not contain a configuration for the live (production) environment. Only two key people in the operations department have access to the storage access keys and SQL credentials for the production environment, which makes it impossible for anyone else to use production storage accidentally during testing.

Figure 4 summarizes the approach Adatum uses for managing configuration and deployment.

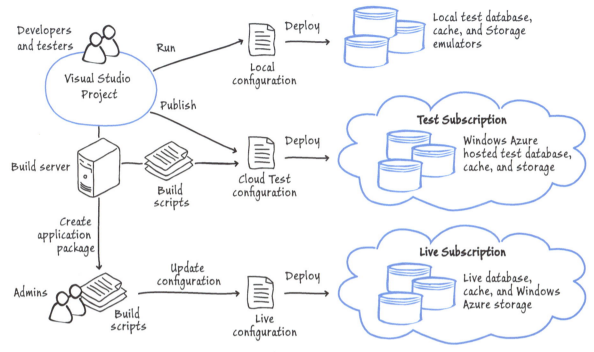

FIGURE 4
Overview of the configuration and deployment approach Adatum uses

Developers and testers can run the application in the local compute and storage emulators, and publish it directly to the test environment using the Publish wizard in Visual Studio. They can also deploy it by using a script that updates the test configuration file, creates the package, and deploys the application to the test environment.

When testing is complete and the application is ready to go live, the administrators who have permission to deploy to the live environment run a series of scripts that automate updating the configuration, package creation, and deployment. The scripts create the deployment package (which is the same for both the test and live environments) and then update the configuration so that it includes the storage account and SQL database settings that allow the application to use the live storage and database. The scripts then deploy the package and the updated configuration to the live production environment.

We use the command-line utility on the build server to automate the generation of our deployment package and configuration file. For testing on the compute emulator or in the test environment we use the Publish option in Visual Studio that packages and deploys the application in a single step.

Preparing for Deployment to Windows Azure

There are a number of steps that Adatum performed to prepare the environments in both the test and production subscriptions before deploying the aExpense application for the first time. Initially, Adatum performed these steps using the Windows Azure Management Portal. However, it plans to automate many of these steps, along with the deployment itself, using the Windows Azure PowerShell cmdlets.

Before deploying the aExpense application to a subscription, Adatum created a Cloud Service to host the application. The Cloud Service determines the Windows Azure datacenter where Adatum will deploy the application and the URL where users will access the application.

The aExpense application uses an HTTPS endpoint and therefore requires a certificate to use for SSL encryption. Adatum uploaded a certificate to the Cloud Service in the portal. For more information about how to use HTTPS in Cloud Services, see *Configuring SSL for an Application in Windows Azure.*

Developers, testers, and operations staff can use the Windows Azure Management Portal to configure and deploy the aExpense application, and to access the portal to perform these steps they use Adatum's Microsoft account credentials. However, Adatum scripts many of these operations using the Windows Azure PowerShell cmdlets. For information about how to install and configure the Windows Azure PowerShell cmdlets to use with your subscription, see *Getting Started with Windows Azure PowerShell.*

> *The Windows Azure PowerShell cmdlets use the Windows Azure Service Management REST API to communicate with Windows Azure. The communication is secured with a management certificate, which is downloaded and installed on the client machine as part of the Windows Azure PowerShell cmdlets installation. This means you are not prompted for credentials when you use these cmdlets.*
>
> *If you use the Publish Windows Azure Application wizard in Visual Studio to deploy directly to Windows Azure, it also uses a management certificate from Windows Azure. For more information, see Publishing Windows Azure Applications to Windows Azure from Visual Studio.*

Deploying to Cloud Services in Windows Azure

When you deploy a Cloud Services application to Windows Azure, you upload the service package and configuration files to a Cloud Service in your Windows Azure subscription, specifying whether you are deploying to the production or staging environment within that Cloud Service.

To help troubleshoot deployments, you can enable Remote Desktop Access to the role when you perform the deployment.

You can deploy an application by uploading the files using the Windows Azure Management Portal, by using the Publish Windows Azure Application wizard in Visual Studio, or by using the Windows Azure PowerShell cmdlets. Both the Visual Studio wizard and the PowerShell cmdlets authenticate with your subscription by using a management certificate instead of a Microsoft account.

For deploying the aExpense application to the test subscription, Adatum sometimes uses the wizard in Visual Studio, but typically uses a PowerShell script. Adatum always deploys to the production subscription using a PowerShell script. Adatum also needs to update the Web.config file for each deployment because some values cannot be placed in the appropriate service configuration file (for more details see the section "Application Configuration" earlier in this chapter).

Using Deployment Scripts

Manually modifying your application's configuration files before you deploy the application to Windows Azure is an error-prone exercise that introduces unnecessary risk in a production environment. The developers at Adatum have developed a set of deployment scripts that automatically update the configuration files, package the application files, and upload the application to Windows Azure.

Automating the deployment of applications to Windows Azure using scripts will make it much easier to manage applications running in the cloud.

The automated deployment of the aExpense application in production is handled in two stages. The first stage uses an MSBuild script to compile and package the application for deployment to Windows Azure. This build script uses a custom MSBuild task to edit the configuration files for a cloud deployment, adding the production storage connection details. The second stage uses a Windows PowerShell script with some custom cmdlets to perform the deployment to Windows Azure.

The MSBuild script for the aExpense application uses a custom build task named **RegexReplace** to make the changes during the build. The example shown here replaces the development storage connection strings with the Windows Azure storage connection strings.

You should also have a target that resets the development connection strings for local testing.

XML
```
<Target Name="SetConnectionStrings">
  <RegexReplace
    Pattern='Setting name="Microsoft.WindowsAzure.Plugins
                           .Diagnostics.ConnectionString"
    value="UseDevelopmentStorage=true"'
    Replacement='Setting name="Microsoft.WindowsAzure.Plugins
                               .Diagnostics.ConnectionString"
    value="DefaultEndpointsProtocol=https;
           AccountName=$(StorageAccountName);
           AccountKey=$(StorageAccountKey)"'
    Files='$(AzureProjectPath)\$(ServiceConfigName)'/>
  <RegexReplace
    Pattern='Setting name="DataConnectionString"
    value="UseDevelopmentStorage=true"'
    Replacement='Setting name="DataConnectionString"
    value="DefaultEndpointsProtocol=https;
           AccountName=$(StorageAccountName);
           AccountKey=$(StorageAccountKey)"'
    Files='$(AzureProjectPath)\$(ServiceConfigName)'/>
  <RegexReplace
    Pattern='Setting name="Microsoft.WindowsAzure.Plugins
                          .Caching.ConfigStoreConnectionString"
    value="UseDevelopmentStorage=true"'
    Replacement='Setting name="Microsoft.WindowsAzure.Plugins
                              .Caching.ConfigStoreConnectionString"
    value="DefaultEndpointsProtocol=https;
           AccountName=$(StorageAccountName);
           AccountKey=$(StorageAccountKey)"'
    Files='$(AzureProjectPath)\$(ServiceConfigName)'/>
  <RegexReplace
    Pattern='connectionString="Data Source=LocalTestSQLServer;
           Initial Catalog=aExpense;Integrated Security=True"'
    Replacement='connectionString="Data Source=
                                  $(DatabaseServer);
                                  Initial Catalog=$(DatabaseName);
                                  UId=$(UserName);
                                  Pwd=$(Password);"'
    Files='$(WebProjectConfig)'/>
</Target>
```

*The source code for the **RegexReplace** custom build task is available in the download for this phase. Note that the values used in the example scripts do not exactly match those shown above.*

The team at Adatum then developed a Windows PowerShell script (**deploy.ps1**) that will deploy the packaged application to Windows Azure, and can invoke this script from an MSBuild task. The script needs the following pieces of information to connect to Windows Azure. You must replace the values for thumbprint and subscription ID with the values for your own Windows Azure account:

- **Build path**. This parameter identifies the folder where you build your deployment package. For example: C:\AExpenseBuildPath.
- **Package name**. This parameter identifies the package to upload to Windows Azure. For example: aExpense.Azure.cspkg.
- **Service config**. This parameter identifies the service configuration file for your project. For example: ServiceConfiguration.cscfg.
- **Service name**. This parameter identifies the name of your Windows Azure hosted service. For example: aExpense.
- **Thumbprint**. This is the thumbprint of the service management certificate.
- **Subscription ID (sub)**. This parameter must be set to the name of your Windows Azure subscription. You can find your Subscription ID in the properties pane of the Management Certificates page in the original Windows Azure Management Portal.
- **Slot**. This parameter identifies the environment were you will deploy (Production or Staging)
- **Storage account key (storage)**. This parameter is the key for your storage account, which can be obtained from the Windows Azure Management Portal.

The scripts described in this section use a Windows Azure Management Certificate to authenticate with the Windows Azure subscription. This certificate was installed from the publishing settings downloaded when the operator at Adatum installed Windows Azure PowerShell cmdlets. By default, this certificate is stored in the personal certificate store of the person who installed the cmdlets. You should ensure that only authorized users have access to this certificate because it grants full access to the Windows Azure subscription it came from. This is not the same certificate as the SSL certificate used by the HTTPS endpoint in the aExpense application.

Script

```
$buildPath = $args[0]
$packagename = $args[1]
$serviceconfig = $args[2]
$servicename = $args[3]
$mgmtcertthumbprint = $args[4]
$cert = Get-Item cert:\CurrentUser\My\$mgmtcertthumbprint
$sub = $args[5]
$slot = $args[6]
$storage = $args[7]
$package = join-path $buildPath $packageName
$config = join-path $buildPath $serviceconfig
$a = Get-Date
$buildLabel = $a.ToShortDateString() + "-" + $a.ToShortTimeString()

#Important!  When using file based packages (non-http paths),
#the cmdlets will attempt to upload the package to blob storage
#for you automatically.  If you do not specify a
#-StorageServiceName option, it will attempt to upload a
#storage account with the same name as $servicename.  If that
#account does not exist, it will fail.  This only applies to
#file-based package paths.

#Check for 32-bit or 64-bit operating system
$env = Get-Item Env:\ProgramFiles*x86*
if ($env -ne $null) {
  $PFilesPath = $env.value
} else {
  $env = Get-Item Env:\ProgramFiles
  $PFilesPath = $env.value
}

$ImportModulePath = Join-Path $PFilesPath "Microsoft SDKs\Windows Azure\PowerShell\Microsoft.
WindowsAzure.Management.psd1"
Import-Module $ImportModulePath

Set-AzureSubscription -SubscriptionName Adatum -Certificate $cert
        -SubscriptionId $sub -CurrentStorageAccount $storage

$hostedService = Get-AzureService $servicename |
              Get-AzureDeployment -Slot $slot

if ($hostedService.Status -ne $null)
```

```
{
  $hostedService |
    Set-AzureDeployment -Status -NewStatus "Suspended"
  $hostedService |
    Remove-AzureDeployment -Force
}
Get-AzureService -ServiceName $servicename |
  New-AzureDeployment -Slot $slot -Package $package
        -Configuration $config -Label $buildLabel

Get-AzureService -ServiceName $servicename |
    Get-AzureDeployment -Slot $slot |
    Set-AzureDeployment -Status -NewStatus "Running"
```

The script first verifies that the Windows Azure PowerShell cmdlets are loaded. Then, if there is an existing service, the script suspends and removes it. The script then deploys and starts the new version of the service.

MSBuild can invoke the Windows PowerShell script in a task and pass all the necessary parameter values:

The examples here deploy aExpense to the staging environment. You can easily modify the scripts to deploy to production. You can also script in-place upgrades when you have multiple role instances.

XML
```xml
<Target Name="Deploy"
  DependsOnTargets="SetConnectionStrings;Build;DeplotCert">
  <MSBuild
    Projects="$(AzureProjectPath)\$(AzureProjectName)"
    Targets="CorePublish"
    Properties="Configuration=$(BuildType)"/>
  <Exec WorkingDirectory="$(MSBuildProjectDirectory)"
    Command=
    "$(windir)\system32\WindowsPowerShell\v1.0\powershell.exe
    -NoProfile -f deploy.ps1 $(PackageLocation) $(PackageName)
    $(ServiceConfigName) $(HostedServiceName)
    $(ApiCertThumbprint) $(SubscriptionKey) $(HostSlot)
    $(StorageAccountName)" />
</Target>
```

> *See the release notes provided with the examples for information on using the Windows PowerShell scripts.*

The aExpense application uses an HTTPS endpoint, so as part of the automatic deployment, Adatum needed to upload the necessary certificate. The developers created a Windows PowerShell script named **deploycert.ps1** that performs this operation.

This script needs the following pieces of information to connect to Windows Azure. You must specify the values for thumbprint and subscription ID with the values for your own Windows Azure account:

- **Service name**. This parameter identifies the name of your Windows Azure hosted service. For example: aExpense.

- **Certificate to deploy**: This parameter specifies the certificate you will deploy. It's the one the application is using. Specify the full path to the **.pfx** file that contains the certificate

- **Certificate password**: This parameter specifies password of the certificate you will deploy.

- Thumbprint. This is the thumbprint of the service management certificate.

- **Subscription ID (sub)**. This parameter must be set to the name of your Windows Azure subscription. You can find your Subscription ID in the properties pane of the Management Certificates page in the original Windows Azure Management Portal.

- **Algorithm**: This parameter specifies the algorithm used to create the certificate thumbprint

- **Certificate thumbprint**: This parameter must be set to the value of the thumbprint for the certificate in the **.pfx** file that you will deploy.

```
Script
$servicename = $args[0]
$certToDeploy = $args[1]
$certPassword = $args[2]
$mgmtcertthumbprint = $args[3]
$cert = Get-Item cert:\CurrentUser\My\$mgmtcertthumbprint
$sub = $args[4]
$algo = $args[5]
$certThumb = $args[6]

$env = Get-Item Env:\ProgramFiles*x86*
if ($env -ne $null)
{
  $PFilesPath = $env.value
}
else
{
  $env = Get-Item Env:\ProgramFiles
  $PFilesPath = $env.value
}
```

```
$ImportModulePath = Join-Path $PFilesPath "Microsoft SDKs\Windows Azure\PowerShell\Microsoft.
WindowsAzure.Management.psd1"
Import-Module $ImportModulePath

Set-AzureSubscription -SubscriptionName Adatum -Certificate $cert
                      -SubscriptionId $sub

try
{
  Remove-AzureCertificate -ServiceName $servicename
          -ThumbprintAlgorithm $algo -Thumbprint $certThumb
}
catch {}

Add-AzureCertificate -ServiceName $servicename -CertToDeploy $certToDeploy -Password $certPassword
```

An MSBuild file can invoke this script and pass the necessary parameters. The following code is an example target from an MSBuild file.

```xml
XML
<Target Name="DeployCert">
  <Exec WorkingDirectory="$(MSBuildProjectDirectory)"
    Command=
      "$(windir)\system32\WindowsPowerShell\v1.0\powershell.exe
      -f deploycert.ps1 $(HostedServiceName) $(CertLocation)
      $(CertPassword) $(ApiCertThumbprint) $(SubscriptionKey)
      $(DeployCertAlgorithm) $(DeployCertThumbprint)" />
</Target>
```

Continuous Delivery

Adatum's development team use Microsoft Team Foundation Server (TFS) to manage development projects. Windows Azure can integrate with both an on-premises TFS installation and with the Windows Azure hosted Team Foundation Services to automate deployment when each check-in occurs. Adatum is considering using this feature to automate deployment to the test environment.

The integration provides a package build process that is equivalent to the Package command in Visual Studio, and a publishing process that is equivalent to the Publish command in Visual Studio. This would allow developers to automatically create packages and deploy them to Windows Azure after every code check-in.

> For more information about implementing Continuous Delivery using TFS, see "Continuous Delivery for Cloud Applications in Windows Azure" and "Announcing Continuous Deployment to Azure with Team Foundation Service."

Using a Mock Issuer

By default, the downloadable version of aExpense is set up to run on a standalone development workstation. This is similar to the way you might develop your own applications. It's generally easier to start with a single development computer.

To make this work, the developers of aExpense wrote a small stub implementation of an issuer. You can find this code in the downloadable Visual Studio solution. The project is in the **Dependencies** folder and is named **Adatum.SimulatedIssuer**.

When you first run the aExpense application, you'll find that it communicates with the stand-in issuer. The issuer issues predetermined claims. It's not very difficult to write this type of component, and you can reuse the downloadable sample, or you can start with the template included in the Windows Identity Foundation (WIF) SDK.

> *You can download the WIF SDK from the Microsoft Download Center. The guide "A Guide to Claims-Based Identity and Access Control" describes several ways you can create a claims issuer.*

Converting to a Production Issuer

When you are ready to deploy to a production environment, you'll need to migrate from the simulated issuer that runs on your development workstation to a component such as ADFS 2.0.

Making this change requires two steps. First, you need to configure the issuer so that it recognizes requests from your Web application and provides the appropriate claims. You need do this only once unless you change the claims required by the application.

Then, each time you deploy the solution from test to production, you need to modify the Web application's Web.config file using the **FedUtil** utility such that it points to the production issuer. You may be able to automate this change by using deployment scripts, or by adding code that copies values from your service configuration files to the live web.config file at runtime as described in the section "Application Configuration" earlier in this chapter.

> *To learn more about **FedUtil** and configuring applications to issue claims, take a look at the guide "A Guide to Claims-Based Identity and Access Control."*

You can refer to documentation provided by your production issuer for instructions about how to add a relying party and how to add claims rules.

When you forward a request to a claim issuer, you must include a *wreply* parameter that tells the claim issuer to return the claims. If you are testing your application locally and in the cloud, you don't want to hard code this URL because it must reflect the real address of the application. The following code shows how the aExpense application generates the *wreply* value dynamically in the Global.asax.cs file.

*Building the **wreply** parameter dynamically simplifies testing the application in different environments.*

```csharp
C#
private void
  WSFederationAuthenticationModule_RedirectingToIdentityProvider
  (object sender, RedirectingToIdentityProviderEventArgs e)
{
    // In the Windows Azure environment, build a wreply parameter
    // for  the SignIn request that reflects the real
    // address of the application.
    HttpRequest request = HttpContext.Current.Request;
    Uri requestUrl = request.Url;
    StringBuilder wreply = new StringBuilder();

    wreply.Append(requestUrl.Scheme); // e.g. "http" or "https"
    wreply.Append("://");
    wreply.Append(request.Headers["Host"] ??
        requestUrl.Authority);
    wreply.Append(request.ApplicationPath);

    if (!request.ApplicationPath.EndsWith("/"))
    {
        wreply.Append("/");
    }

    e.SignInRequestMessage.Reply = wreply.ToString();
}
```

Accessing Diagnostics Log Files

The aExpense application uses the Logging Application Block and the Exception Handling Application Block to capture information from the application and write it to the Windows event log. The Cloud Services version of the application continues to use the same application blocks, and through a configuration change, it is able to write log data to the Windows Azure logging system.

The Logging Application Block and the Exception Handling Application Block are part of the Enterprise Library. We use them in a number of applications within Adatum.

For aExpense to write logging information to Windows Azure logs, Adatum made a change to the Web.config file to make the Logging Application Block use the Windows Azure trace listener.

```xml
XML
<listeners>
<add listenerDataType="Microsoft.Practices.EnterpriseLibrary.
    Logging.Configuration.SystemDiagnosticsTraceListenerData,
    Microsoft.Practices.EnterpriseLibrary.Logging,
    Version=5.0.414.0, Culture=neutral,
    PublicKeyToken=31bf3856ad364e35"
  type="Microsoft.WindowsAzure.Diagnostics
        .DiagnosticMonitorTraceListener,
        Microsoft.WindowsAzure.Diagnostics, Version=1.0.0.0,
        Culture=neutral,PublicKeyToken=31bf3856ad364e35"
  traceOutputOptions="Timestamp, ProcessId"
  name="System Diagnostics Trace Listener" />
</listeners>
```

If you create a new Windows Azure Project in Visual Studio, the Web.config file will contain the configuration for the Windows Azure trace listener. The following code example from the Web.config file shows the trace listener configuration you must add if you are migrating an existing ASP.NET web application.

```xml
XML
<system.diagnostics>
    <trace>
      <listeners>
        <add type="Microsoft.WindowsAzure.Diagnostics
            .DiagnosticMonitorTraceListener,
            Microsoft.WindowsAzure.Diagnostics, Version=1.0.0.0,
            Culture=neutral, PublicKeyToken=31bf3856ad364e35"
            name="AzureDiagnostics">
          <filter type="" />
        </add>
      </listeners>
    </trace>
  </system.diagnostics>
```

By default in Windows Azure, diagnostic data is not automatically persisted to storage; instead, it is held in a memory buffer. In order to access the diagnostic data, you must either add some code to your application that transfers the data to Windows Azure storage, or add a diagnostics configuration file to your project. You can either schedule Windows Azure to transfer log data to storage at timed intervals, or perform this task on-demand.

Adatum decided to use a diagnostics configuration file to control how the log data is transferred to persistent storage; the advantage of using a configuration file is that it enables Adatum to collect trace data from the **Application_Start** method where the aExpense application performs its initialization routines. The following snippet shows the sample diagnostics.wadcfg file from the solution.

```XML
<?xml version="1.0" encoding="utf-8" ?>
<DiagnosticMonitorConfiguration
    xmlns="http://schemas.microsoft.com/ServiceHosting/
          2010/10/DiagnosticsConfiguration"
      configurationChangePollInterval="PT1M"
      overallQuotaInMB="5120">
  <DiagnosticInfrastructureLogs bufferQuotaInMB="1024"
    scheduledTransferLogLevelFilter="Verbose"
    scheduledTransferPeriod="PT1M" />
  <Logs bufferQuotaInMB="1024"
    scheduledTransferLogLevelFilter="Verbose"
    scheduledTransferPeriod="PT1M" />
  <Directories bufferQuotaInMB="1024"
    scheduledTransferPeriod="PT1M">

    <!-- These three elements specify the special directories
         that are set up for the log types -->
    <CrashDumps container="wad-crash-dumps"
             directoryQuotaInMB="256" />
    <FailedRequestLogs container="wad-frq"
             directoryQuotaInMB="256" />
    <IISLogs container="wad-iis" directoryQuotaInMB="256" />

  </Directories>
  <PerformanceCounters bufferQuotaInMB="512"
                    scheduledTransferPeriod="PT1M">
    <!-- The counter specifier is in the same format as the
         imperative diagnostics configuration API -->
    <PerformanceCounterConfiguration
      counterSpecifier="\Processor(_Total)\% Processor Time"
      sampleRate="PT5S" />
  </PerformanceCounters>
  <WindowsEventLog bufferQuotaInMB="512"
    scheduledTransferLogLevelFilter="Verbose"
    scheduledTransferPeriod="PT1M">
    <!-- The event log name is in the same format as the
         imperative diagnostics configuration API -->
    <DataSource name="System!*" />
  </WindowsEventLog>
</DiagnosticMonitorConfiguration>
```

The value of the **overallQuotaInMB** must be more than the sum of the **bufferQuotaInMB** values in the diagnostics.wadcfg file, and you must configure a local storage resource in the Web role named "Diagnostics-Store" that is at least the size of the **overallQuotaInMB** value in the diagnostics configuration file. In this example, the log files are transferred to storage every minute and you can then view them with any storage browsing tool such as the Server Explorer window in Visual Studio.

> *You must also configure the trace listener for the Windows Azure IISHost process separately in an App.config file. For more detailed information about using Windows Azure diagnostics, see Collecting Logging Data by Using Windows Azure Diagnostics on MSDN. The blog post "Configuring WAD via the diagnostics. wadcfg Config File" also contains some useful tips.*

As an alternative to using the diagnostics.wadcfg file Adatum could have used code in the OnStart event of the role. However, this means that changes to the configuration will require the application to be redeployed.

MORE INFORMATION

All links in this book are accessible from the book's online bibliography available at: *http://msdn.microsoft.com/en-us/library/ff803373.aspx*.

The *Windows Azure Developer Center* contains links to plenty of resources to help you learn about developing applications for Windows Azure.

MSDN is a good starting point for learning more about *Windows Azure and Windows Azure SQL Database*.

You can download the latest versions of Windows Azure tools for developing applications using .NET and other languages from the *Windows Azure Developer Center* Downloads page.

"Managing Windows Azure SQL Database using SQL Server Management Studio" contains steps for connecting to and managing Windows Azure SQL Database by using an on-premises instance of SQL Server Management Studio.

"About the Service Management API" contains an overview of managing Windows Azure services using scripts and code.

You can download the Windows Azure PowerShell cmdlets and other management tools from the *developer tools page*.

For more information about caching and the ASP.NET 4 Caching Providers for Windows Azure, see *"Caching in Windows Azure."*

For more information about the Universal providers for profile data, see *"Microsoft ASP.NET Universal Providers"* and the Hands-on Labs that accompany this guide.

Because persisting diagnostic data to Windows Azure storage costs money, we will need to plan how long to keep the diagnostic data in Windows Azure and how we are going to download it for offline analysis.

4

Moving to Windows Azure SQL Database

This chapter describes the third step in Adatum's migration of the aExpense application to Windows Azure. For the initial migration of the aExpense application to the cloud, Adatum chose the IaaS model because it required the fewest modifications to the existing application. In the previous chapter, you saw why and how Adatum decided to move the aExpense web application to the PaaS deployment model by using Windows Azure Cloud Services instead of Windows Azure Virtual Machines. In this chapter you will see how Adatum reviewed its choice of data store, and moved from using SQL Server hosted in a Windows Azure virtual machine to using Windows Azure SQL Database.

The Premise

At present the aExpense application's data store still uses the IaaS model in the form of SQL Server installed in a Windows Azure virtual machine. However, the Windows Azure platform also includes cloud services that offer both relational database capabilities and less structured storage capabilities using the PaaS model. Adatum wants to explore the capabilities and the consequent impact on their migration strategy for these alternative data stores.

Goals and Requirements

In this step along its migration path Adatum wants to evaluate whether it can reduce operating and hosting costs for the data storage functions of the aExpense application. The goals are to reduce the direct hosting costs if this is possible, as well as minimizing the amount of management and maintenance required by Adatum's operators and administrators.

Adatum must first ensure that the data store they choose at this stage is suitable for use with the aExpense application, and it does not have limitations that may prevent the operation or future development of this or other related applications.

When you evaluate the choice of data storage mechanism, remember that you are making a decision that could affect more than just the application you are currently migrating. If you plan to move additional on-premises applications to the cloud in the future, or create new applications that share the same data, you should consider how these will be affected by your choice of data storage mechanism.

Adatum would like to be able to change just the connection strings that define how the application connects to the data store, and continue to use the same SQL Server tools to manage and interact with its data. However, Adatum must consider how management tasks such as backing up the data can be accomplished when it moves away from using a hosted instance of SQL Server.

Finally, Adatum must ensure that the solution it chooses provides an acceptable level of performance, reliable connectivity, and sufficient data storage and throughput capabilities.

OVERVIEW OF THE SOLUTION

This section of the chapter describes how Adatum evaluated the Windows Azure PaaS services for data storage, and then compares Windows Azure SQL Database to SQL Server. It also describes how Adatum explored the limitations of SQL Database in terms of its impact on the aExpense application.

PaaS Options for Data Storage

Windows Azure PaaS approach offers these two built-in mechanisms for storing data:

- **Windows Azure SQL Database**. This is a highly available and scalable cloud database service built on SQL Server technologies, and it supports the familiar T-SQL based relational database model. It can be used with applications hosted in Windows Azure, and with other applications running on-premises or hosted elsewhere.

- **Windows Azure Storage**. This provides the following core services for persistent and durable data storage in the cloud. The services support both managed and REST APIs that can be used from within Azure-hosted or on-premises applications.

  - **Windows Azure tables** provide a non-relational table-structured storage mechanism. Tables are collections of entities that do not have an enforced schema, which means a single table can contain entities that have different sets of properties. This mechanism is primarily aimed at scenarios where large volumes of data must be stored, while being easy to access and update.

  - **Windows Azure blobs** provide a series of containers aimed at storing text or binary data. Block blob containers are ideal for streaming data, while page blob containers can be used for random read/write operations.

  - **Windows Azure queues** provide a mechanism for reliable, persistent messaging between role instances, such as between a web role and a worker role. They can also be used between roles and non-hosted applications and services, though Windows Azure Service Bus messaging is typically more suited to this scenario.

  - **Windows Azure drives** provide a mechanism for applications to mount a single volume NTFS VHD as a page blob, and access it as though it were a local hard drive. They are aimed at scenarios where applications rely on access to a physical file system.

Adatum previously discounted using Windows Azure storage during the early steps of the migration due to the requirement to change the data access mechanism to work with the table and blob storage APIs. At this stage, Adatum wants to be able to continue using standard SQL Server tools, and change only the database connection strings. Therefore, the logical choice for this step is to use Windows Azure SQL Database.

Windows Azure SQL Database is SQL Server for the cloud. Therefore, Adatum should be able to move the aExpense data from SQL Server running in a Windows Azure Virtual Machine to SQL Database without requiring changes to the way that the application works.

Using Windows Azure SQL Database provides a compelling option for storing data because it removes the requirement to manage the operating system and database software. Developers can simply migrate their application data to the Windows Azure SQL Database hosted database service. They can use the Management Portal to configure the database, create and manage the tables, run scripts, and monitor the database. In addition they can use SQL Server Management Console, Visual Studio, scripts, or any other compatible tools to migrate and interact with the database.

For the majority of applications, moving from SQL Server to SQL Database does not require any changes in either the database or the application code other than modifying the connection strings. However, there are some differences in the feature set supported by SQL Database as compared to SQL Server, which you must evaluate before making your decision.

> The MSDN topic. **"Migrating Databases to Windows Azure SQL Database (formerly SQL Azure)"** provides an overview of how to migrate a database to Windows Azure SQL Database. You can also use the **Migration Wizard** to help you to migrate your local SQL Server databases to Windows Azure SQL Database.

Comparing SQL Server and Windows Azure SQL Database

Given the similarities between SQL Server and Windows Azure SQL Database, why should Adatum choose one over the other? The key to this choice is to understand the advantages and limitations of the IaaS and PaaS models. In the cloud, SQL Server is available if you opt for the IaaS model while and SQL Database is available if you opt for the PaaS model.

If you choose to run SQL Server on a virtual machine in the cloud, you are responsible for managing both the operating system and the SQL Server installation; and ensuring that they are kept up to date with patches, that they are securely configured, and so forth. Windows Azure keeps copies of VMs for resilience and you can choose to have the storage account that holds these copies geo-replicated across different datacenters, but you do pay for the blob storage that holds these copies.

While SQL Server and Windows Azure SQL Database are fundamentally similar, there are some important differences that you must be aware of when making a decision on how to store your applications' data.

If you choose SQL Database, Microsoft handles all of this management for you. However, taking the PaaS route does mean that you give up some level of control over your environment.

In terms of cost, the situation is more complex because you have three basic choices for a SQL data store:

- **License SQL Server yourself, then install and run it in a Windows Azure Virtual Machine**. In this scenario, you may be able to reuse an existing license, but the SQL Server costs are all upfront and you must try to ensure at the start of your project that you choose and license the best version of SQL Server for your application.

- **Use a Windows Azure virtual machine image with SQL Server pre-installed**. In this scenario, you pay by the hour for using SQL Server; you can stop at any time, and move to a larger or smaller virtual machine at any time. This is still an IaaS solution because you are responsible for managing the operating system and the SQL Server installation.

- **Use Windows Azure SQL Database**. In this scenario, you also have the flexibility of the pay-as-you-go pricing model, but you now have a PaaS model for deploying a SQL data store in Windows Azure. After you create a logical SQL Database server instance you can create multiple databases easily and quickly without needing to manage the server itself. Windows Azure automatically manages allocation of databases across all the physical servers.

See Chapter 6, "Evaluating Cloud Hosting Costs" for more detailed information about the relative costs of these options.

Limitations of Windows Azure SQL Database

Windows Azure SQL Database does have some limitations compared to SQL Server. Developers will not be able to use free text search, some XML handling capabilities, procedures that require common language runtime (CLR) programmability, scheduled tasks, and distributed queries that rely on the SQL Service Broker. In addition, some data types (including user-defined CLR types) are not supported.

However, Windows Azure SQL Database does support stored procedures and the majority of Transact-SQL functions and operators, and you may also be able to implement workarounds for some of the limitations, such as using a separate mechanism such as the *Quartz scheduler* in a Cloud Service role to drive scheduled tasks.

SQL Server may also provide better performance under certain high load conditions and complex query execution than Windows Azure SQL Database, which can be subject to throttling under very high load conditions. However, SQL Database supports a mechanism called federations that allows you to partition data by a specific key across multiple databases while still being able to interact with the data using standard queries.

> See "Federations in Windows Azure SQL Database" for more information about using SQL Database federations.

One other major difference is that Windows Azure SQL Database automatically replicates the data to multiple datacenters to provide resilience against failure. However, this is not a complete backup solution because it does not provide access to previous versions of the database. A database backup routine must be established for critical data when using both a hosted SQL Server and Windows Azure SQL Database. The section "Data Management" later in this chapter provides an example of a simple backup technique.

> For more details of these differences, see "Windows Azure SQL Database Overview" and "Guidelines and Limitations (Windows Azure SQL Database)".

After considering all of the options available, and their impact on the migration of the aExpense and other applications, the developers decided that the simplicity offered by Windows Azure SQL Database was the most attractive. The few limitations it imposes will not prevent them from using it with the existing data and procedures in the aExpense application, and in other applications that Adatum intends to create in the future.

Database Management and Data Backup

Windows Azure SQL Database exposes remote management connectivity that is compatible with SQL Server management tools and other third party tools. Administrators, operators, developers, and testers can use SQL Server Management Console, Visual Studio, scripts, or any other compatible tools to connect to a SQL Database server instance in the cloud to migrate data, perform management tasks, and to and interact with the database and the data it holds. This familiar approach makes changing from SQL Server to SQL Database a relatively pain-free process.

Windows Azure SQL Database automatically maintains three replicas of database in different locations within the same datacenter to protect against hardware failure that may affect an individual node within the datacenter. However, it's important to recognize that this is not a backup solution in the same sense that administrators and operators perform backup procedures for an on-premises or cloud-hosted SQL Server database.

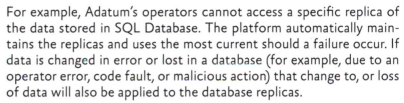

For example, Adatum's operators cannot access a specific replica of the data stored in SQL Database. The platform automatically maintains the replicas and uses the most current should a failure occur. If data is changed in error or lost in a database (for example, due to an operator error, code fault, or malicious action) that change to, or loss of data will also be applied to the database replicas.

Therefore it is vital that Adatum adopts a backup strategy that works with SQL Database. For example, Adatum may choose to back up the data to an on-premises database or a separate cloud-hosted database by using replication or the Windows Azure database import and export feature; by using the SQL Server Management Studio to create a SQL Server Backup for the data; or by using third party tools to copy the data between Windows Azure SQL Database and blob storage. These tools can also be used for initializing the data in a database from a backup copy.

> See "Windows Azure SQL Database Backup and Restore" and "Business Continuity in Windows Azure SQL Database" for information about how you can back up and restore Windows Azure SQL Databases.

Windows Azure SQL Database automatically maintains copies of your database, but you cannot depend on this as a backup strategy. For example, if you accidently delete data in your current database, it will be removed from all the replica copies as well.

Database Connection Reliability

Adatum has noticed that the application running in the cloud very occasionally suffers from a failed connection to the database. While the application was running on premises these kinds of events were very rare, mainly due to the high speed physical connection between the application servers and the database servers within Adatum's datacenter. When the application was moved to cloud, the data was deployed in a hosted SQL Server running in a Windows Azure Virtual Machine, and so connectivity between the application and the database was no longer over a dedicated network managed by Adatum.

While the connectivity within the Windows Azure datacenters is much faster and more reliable than is available on the Internet, Adatum can no longer manage traffic allocation and so occasional connection failures may occur. In addition, when Adatum moves from using SQL Server to the shared data storage mechanism that is Windows Azure SQL Database, there are additional factors to consider. For example, Windows Azure SQL Database will attempt to connect for thirty seconds, but may fail when the database encounters severe load conditions, or if the server is automatically recycled following a failure.

In the cloud, the types of error that often disappear if you retry the operation are known as transient faults. They can occur for many reasons, including intermittent connectivity issues within the cloud or throttling behavior in the cloud service.

Adatum must address the issue of how to handle these kinds of transient faults in its applications' data access code. If the timeout expires before establishing a connection, or if a connection to Windows Azure SQL Database drops while an application is using the connection, an error occurs; which the application must handle. How the application handles these errors depends on the specific circumstances, but possible strategies include immediately trying to re-establish the connection, keep retrying the connection until it succeeds, report the error to the user, or log the error and move on to another task.

If a database transaction was in progress at the time of the failure, the application should retry the transaction. It is possible for a connection to fail between sending a message to commit a transaction and receiving a message that reports the outcome of the transaction. In this circumstance, the application must have some way of checking whether the transaction completed successfully in order to determine whether to retry it.

Whereas Adatum wanted to avoid making unnecessary changes to the application code in previous migration steps, the developers have realized that they must resolve these issues now that Adatum has chosen to use Windows Azure SQL Database by implementing retry logic in the aExpense application.

Implementing Retry Logic for Database Connections

It is possible to create your own logic for retrying failed or dropped connections by writing a suitable delegate. The blog post *"SaveChangesWithRetries and Batch Option"* describes how to use the **Retry-Policy** delegate in the **Microsoft.WindowsAzure.StorageClient** namespace for this. The post describes using this delegate to retry saving changes to Windows Azure table storage, but you could adapt this approach to work with a context object in LINQ to SQL or ADO.NET Entity Framework.

However, a simpler option is to use the *Transient Fault Handing Application Block* that is part of the patterns & practices Enterprise Library. This application block exposes methods that allow you to execute code that connects to a wide range of Windows Azure services including SQL Database, Windows Azure Service Bus, Windows Azure Storage, and Windows Azure Caching Service.

You can configure a range of different retry policies, and write code to handle the success and failure events for the connection. You can easily add the Transient Fault Handling Application Block to an existing application; it requires very few code changes in your application, and you can customize its behavior to meet your specific requirements through configuration. Customization options include the choice of retry strategy to use in different circumstances; for example, you can choose between fixed interval, incremental interval, and exponential interval retry strategies.

This is the solution that the developers at Adatum chose to adopt.

> For more information see *"The Transient Fault Handling Application Block."* The block is extensible, which enables you to add support other for other cloud services. You can find guidance on how to extend the block in *"Extending and Modifying the Transient Fault Handling Application Block."*

INSIDE THE IMPLEMENTATION

The Hands-on Labs that accompany this guide provide a step-by-step walkthrough of parts of the implementation tasks Adatum carried out on the aExpense application at this stage of the migration process.

Now is a good time to walk through this step in the process of migrating aExpense into a cloud-based application in more detail. As you go through this section, you may want to download the Visual Studio solution from *http://wag.codeplex.com/*. This solution contains an implementation of the aExpense application (in the **Azure-CloudService-WADB** folder) after the migration step described in this chapter. If you are not interested in the mechanics, you should skip to the next section.

Connecting to Windows Azure SQL Database

Connecting to Windows Azure SQL Database instead of SQL Server requires only a configuration change.

In all of the previous migration steps, Adatum used a hosted SQL Server database to store the data for the aExpense application. In this phase, the team moved the database to Windows Azure SQL Database and the data access code in the application remained unchanged. The only thing that needs to change is the connection string in the configuration file.

```XML
<add name="aExpense" connectionString=
  "Data Source={Server Name};
   Initial Catalog=aExpense;
   UId={User Id};
   Pwd={User Password};
   Encrypt=True;
   TrustServerCertificate=False;
   Connection Timeout=30;"
  providerName="System.Data.SqlClient" />
```

*The values of **Server Name**, **User Id**, and **User Password** are specific to your Windows Azure SQL Database account.*

There are two things to notice about the connection string. First, notice that, because Windows Azure SQL Database does not support Windows Authentication, the credentials for your Windows Azure SQL Database account are stored in plain text. You should consider encrypting this section of the configuration file as described in Chapter 3, "Moving to Windows Azure Cloud Services."

The second thing to notice about the connection string is that it specifies that all communications with Windows Azure SQL Database are encrypted. Even though your application may reside on a computer in the same data center as Windows Azure SQL Database, you have to treat that connection as if it was using the internet.

Any traffic within the data center is considered to be "Internet" and should be encrypted.

Handling Transient Connection Failures

When you try to connect to Windows Azure SQL Database, you can specify a connection timeout value in your data access connection strings. Windows Azure SQL Database will attempt to connect for thirty seconds, and so you should set the connection timeout in your code to at least this value. Adatum set the connection timeout to thirty seconds in the aExpense application.

> *Windows Azure SQL Database removes idle connections after thirty minutes, and also imposes throttling to protect all users of the service from unnecessary delays. For a detailed explanation of how this works, and how you can optimize your data access connections, see "Windows Azure SQL Database Connection Management."*

In addition, Adatum decided to use the Enterprise Library Transient Fault Handling Application Block to implement a consistent and configurable retry strategy in the aExpense application. The recommended retry interval for Windows Azure SQL Database is ten seconds.

Adatum added a configuration section for the Transient Fault Handling Application Block to the Web.config file for the **aExpense.Azure** project. This configuration defines several retry strategies that the application can use, and identifies the "Fixed Interval Retry Strategy" as the default.

You can also add protection to your Windows Azure SQL Database instance by configuring the firewall in the Management Portal. You can use the firewall to specify the IP addresses of the computers that are permitted to connect to your Windows Azure SQL Database server and control if other Windows Azure services can connect to your SQL Database.

```xml
XML
<configuration>
  <configSections>
    <section name="RetryPolicyConfiguration"
      type="Microsoft....RetryPolicyConfigurationSettings,
            Microsoft.Practices...TransientFaultHandling,
            Version=5.0.1118.0, Culture=neutral,
            PublicKeyToken=31bf3856ad364e35"
            requirePermission="true"/>
    ...
  </configSections>
  <RetryPolicyConfiguration
      defaultRetryStrategy="Fixed Interval Retry Strategy">
    <incremental name="Incremental Retry Strategy"/>
    <fixedInterval name="Fixed Interval Retry Strategy"/>
    <exponentialBackoff name="Exponential Backoff Retry Strategy"
                        maxRetryCount="3"/>
  </RetryPolicyConfiguration>
  <typeRegistrationProvidersConfiguration>
    ...
    <add sectionName="RetryPolicyConfiguration"
        name="RetryPolicyConfiguration"/>
  </typeRegistrationProvidersConfiguration>
  ...
</configuration>
```

For more information about how to configure these strategies, see *"Specifying Retry Strategies in the Configuration"* on MSDN.

> Adatum chose to define their retry strategies in a configuration file rather than in code to make it easier to modify them in the future.

The following code sample from the **ExpenseRepository** class shows how Adatum uses a retry strategy when the application tries to save an expense item to the SQL Database database. The application uses the default retry policy when it invokes the **SubmitChanges** method.

```C#
using Microsoft.Practices.EnterpriseLibrary.WindowsAzure
                            .TransientFaultHandling;
using Microsoft.Practices.TransientFaultHandling;
using Model;

...

private readonly RetryPolicy sqlCommandRetryPolicy;

...

public void SaveExpense(Model.Expense expense)
{
    using (var db = new DataAccessLayer.ExpensesDataContext
        (this.expenseDatabaseConnectionString))
    {
        var entity = expense.ToEntity();
        db.Expenses.InsertOnSubmit(entity);

        foreach (var detail in expense.Details)
        {
            var detailEntity = detail.ToEntity(expense);
            db.ExpenseDetails.InsertOnSubmit(detailEntity);
        }

        this.sqlCommandRetryPolicy.ExecuteAction(
            () => db.SubmitChanges());
    }
}
```

SETUP, DEPLOYMENT, MANAGEMENT, AND MONITORING

This section discusses the way that Adatum manages the setup and deployment of the Windows Azure SQL Database server and the databases it contains, and how administrators and operators at Adatum manage and monitor the database.

Data for Development and Testing

It is possible to create separate databases within your SQL Database server instance for testing and for use during development. Alternatively, as when using SQL Server hosted in a virtual machine, you can use separate server instances to minimize the chances of corrupting live data. Adatum has a separate Windows Azure subscription that is uses for development and testing, as described in Chapter 3 of this guide. The keys for accessing and deploying to the production subscription are not available to testers and developers.

Data Migration

To migrate an existing database schema to Windows Azure SQL Database, you can use SQL Server Management Studio to export the schema as a Transact-SQL script, and then run the script against Windows Azure SQL Database. To move data to Windows Azure SQL Database, you can use SQL Server Integration Service. The SQL Server Migration Wizard is also very useful for migrating both schema and data.

One issue that Adatum faced is that the **aspnet_regsql.exe** tool used to create the tables for the ASP.NET profile provider will fail to run against a Windows Azure SQL Database. For information about this incompatibility and how to work around it, see *"Updated ASP.net scripts for use with Microsoft SQL Azure."*

When the team at Adatum migrated the live application they created scripts that create the required tables in SQL Database, and transfer data from the on-premises version of SQL Server to Windows Azure SQL Database.

> The ASP.NET providers are not designed to retry operations when they encounter a transient fault. Adatum is investigating alternative solutions to storing the profile data for the aExpense application.

> *For more information about data migration and the tools that are available to help, see "Data Migration to Windows Azure SQL Database Tools and Techniques" and "Choosing Tools to Migrate a Database to Windows Azure SQL Database."*

Data Management

To enable backup and restore functionality, Adatum plans to create a copy of the aExpense database daily using the following Transact-SQL command and maintain three rolling copies of the previous three days.

```
T-SQL
CREATE DATABASE aExpense_copy_[date] AS COPY OF [source_server_
name].aExpense
```

This technique creates a copy of the aExpense database that includes all of the transactions that committed up to the time when the copy operation completed. Adatum can restore to a previous version of the database by renaming the copy, such as aExpense_copy_080612 to aExpense. However, Adatum will be charged for the storage required to maintain multiple copies of the database

> This technique does not allow for point-in-time restore operations as you can do with SQL Server by running the RESTORE command.

Adatum also plans to investigate using the SQL Database Import/Export Service to create a backup of the aExpense database in a different data center from the one where it hosts the aExpense application. This will ensure that Adatum has a copy of the aExpense data in the event of a disaster that leads to the data center where the aExpense application is hosted becoming unavailable.

Database Monitoring

SQL Database enables a subset of the dynamic management views that operators can use to diagnose performance problems in the database. Adatum plans to use these views in the aExpense database.

For more information, see *"Monitoring Windows Azure SQL Database Using Dynamic Management Views"* on MSDN.

MORE INFORMATION

All links in this book are accessible from the book's online bibliography available at: *http://msdn.microsoft.com/en-us/library/ff803373.aspx*.

For information about Windows Azure SQL Database, see *"Windows Azure SQL Database Overview."*

TechNet describes the differences between *SQL Server and Windows Azure SQL Database.*

The MSDN topic *"Migrating Databases to Windows Azure SQL Database (formerly SQL Azure)"* provides an overview of how to migrate a database to Windows Azure SQL Database. You can also use the *Migration Wizard* at to help you to migrate your local SQL Server databases to Windows Azure SQL Database.

See *"Windows Azure SQL Database Backup and Restore"* and *"Business Continuity in Windows Azure SQL Database"* for information about how you can back up Windows Azure SQL Databases.

To encrypt your SQL connection string in the Web.config file, you can use the *Pkcs12 Protected Configuration Provider* that you can download from. The sections "Best Practices on Writing Secure Connection Strings for SQL Database" and "Create and deploy an ASP.NET application with Security in mind" in the post *"Windows Azure SQL Database Connection Security"* discuss using this provider.

For more information about retrying database connections by using Enterprise Library, see *"The Transient Fault Handling Application Block."*

"Overview of Options for Migrating Data and Schema to Windows Azure SQL Database" contains useful information about migrating data to Windows Azure SQL Database.

5   Executing Background Tasks

This chapter walks you through the changes in the cloud-based version of the aExpense application that Adatum made when they added support for uploading, storing, and displaying scanned images of receipts. You'll see how the application uses Windows Azure blob storage to store the image data, how the application uses a worker role in Windows Azure to perform background processing tasks on the images, and how the application uses shared access signatures to control access to the images by users. The chapter also introduces a simple set of abstractions that wrap a worker role, in the expectation that the aExpense application will need to perform additional background tasks in the future.

The Premise

During this phase of the project, the team at Adatum turned their attention to the requirement for a background process in the aExpense application that performs some processing on the scanned images of business expense receipts that users upload.

The original on-premises web application enables users to upload images of their business expense receipts, and the application assigns a unique name to each image file before it saves the image to a file share. It then stores the path to the image in the SQL Server database that holds the business expenses data. The application can then retrieve the image related to an expense submission and display it through the user interface (UI) later.

The completed on-premises application also has a background process that processes the images, which is implemented as a Windows service. This process performs two tasks: it compresses the images to preserve disk space, and it generates a thumbnail image. By default, the application's UI displays the thumbnail, but if a user wants to see a more detailed image, it enables viewing the full-sized version of the image.

Goals and Requirements

Adatum's developers must implement the receipt image upload function into the migrated version of the application running in Windows Azure. Adatum has a number of goals for the implementation of the image processing component of the application.

Storage for scanned receipt images will be one of the monthly costs for the application.

We need to protect the privacy of the users of this application.

Firstly, Adatum wants to minimize the storage requirements for the images while maintaining the legibility of the information on the receipts.

Adatum also wants to maintain the responsiveness of the application and minimize the bandwidth required by the UI. A user shouldn't have to wait after uploading an image while any necessary processing takes place, and the application should display image thumbnails with an option to display a full-sized version.

Finally, Adatum wants to maintain the privacy of its employees by making receipts visible only to the employee who submitted them and to the people responsible for approving the expense submission.

Overview of the Solution

The team at Adatum made several significant changes to the implementation of the aExpense application for the Windows Azure based version in this phase.

Storing Receipt Images

The first decision was to select a storage mechanism for the scanned receipt images. Windows Azure storage provides the following core services for persistent and durable data storage in the cloud:

- **Windows Azure blobs** provide a series of containers aimed at storing text or binary data. Block blob containers are ideal for streaming data, while page blob containers can be used for random read/write operations.
- **Windows Azure queues** provide a mechanism for reliable, persistent messaging between role instances, such as between a web role and a worker role, and between roles and non-hosted applications and services.
- **Windows Azure tables** provide a non-relational table-structured storage mechanism. Tables are collections of entities that do not have an enforced schema, which means a single table can contain entities that have different sets of properties. This mechanism is primarily aimed at scenarios where large volumes of data must be stored, while being easy to access and update.
- **Windows Azure drives** provide a mechanism for applications to mount a single volume NTFS VHD as a page blob, and access it as though it were a local hard drive. They are aimed at scenarios where applications rely on access to a physical file system.

The Windows Azure storage services support both a managed API and a REST API that can be used from within Windows Azure-hosted or on-premises (remote) applications.

A simple approach to the application's storage requirements would be to use the Windows Azure storage drive. This would require minimal changes to the code used in the on-premises version because the Windows Azure drive is a simple NTFS volume that you can access by using the standard .NET I/O classes. The major drawback of this approach is that you can write to the Windows Azure drive from only one role instance at a time. Adatum plans to deploy multiple instances of the aExpense web role to ensure high availability and scalability for the application.

While Windows Azure tables could be used to store the binary data for an image, blob storage is better suited to this type of data. A single entity in table storage is limited to 1 MB in size, whereas a block blob could be up to 200 GB in size. Furthermore, it is easy to reference an image file stored in blob storage with an HTTP or HTTPS address. Therefore, the approach adopted by Adatum was to store the image data in Windows Azure block blob storage. Although this approach requires more changes to the code compared to using a Windows Azure storage drive, it is compatible with using multiple role instances.

Background Processing

The second decision was how to implement background processing for tasks in the aExpense application. Adatum could simply execute the background tasks asynchronously within the application's web role. Carrying out tasks asynchronously in the background offloads work from the application's request handling code, which helps to improve the responsiveness of the UI.

However, this does not allow for individual scaling of the background processing code. If the background tasks require additional processing capacity Adatum must increase the number of deployed instances to accommodate this, even if additional UI processing capacity is not required.

The alternative is to create a separate worker role and execute all of the background tasks within this role. Unlike simply emulating a background role by using asynchronous code in the web role, using a worker role allows for individual scaling of the roles. The developers at Adatum realized that it would be better to implement the UI and the background processing as separate roles that could be scaled individually. However, they also realized that running separate worker role instances will increase the hosting costs of the application because each role instance is charged on an hourly basis.

A worker role is the natural way to implement a background process in Windows Azure. You can scale each role type independently.

Adatum should review the cost estimates for the aExpense application now that the application includes a worker role.

We can use one or more Windows Azure storage queues to communicate between roles. In this example we use a single queue to communicate between the web and the worker role.

Therefore, Adatum decided to implement the image processing service by using a worker role in the cloud-based version of aExpense. Most of the code from the existing on-premises version of the application that compresses images and generates thumbnails was simply repackaged in the worker role. What did change was the way that the image processing service identified when it had a new image to process.

Detecting an New Uploaded Image

Receipt images are uploaded to the aExpense application through the UI, which stores them ready for processing. In the original on-premises version, a Windows service uses the FileSystemWatcher class to generate an event whenever the application saves a new image to the file share. The aExpense application then invokes the image processing logic from the OnCreated event handler, as shown in Figure 1.

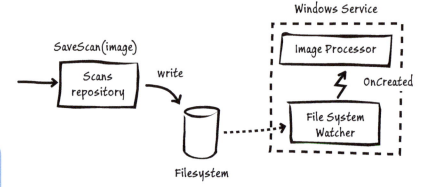

FIGURE 1
On-premises image processing

For the cloud-based version of the application using blob storage, this approach won't work because there is no Windows Azure equivalent of the FileSystemWatcher class for blob storage. Instead, as Figure 2 shows, Adatum decided to use a Windows Azure queue.

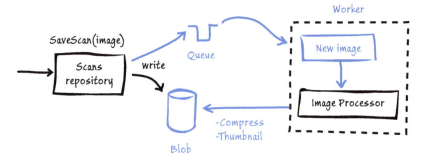

FIGURE 2
Cloud-based image processing

You might consider using Windows Azure Service Bus queues instead of Windows Azure storage queues. Service Bus queues have advantages such as automatic deduplication and long polling timescales, and can provide a level of indirection as well as being accessible through firewalls and network address translation (NAT) routers. However, for simple communication between Cloud Service roles, Windows Azure storage queues typically offer sufficient capabilities and may provide better scalability and performance due to their inherent simplicity when compared to Service Bus queues.

Whenever a user uploads a new image to aExpense as part of an expense submission, the application UI writes a message to the queue and saves the image to blob storage. The worker role will pick up messages from the queue, compress the image, and then generate a thumbnail. The worker role saves the new image and thumbnail in blob storage.

After the worker role completes the image processing, it updates the expense item entity in Windows Azure table storage to include references to the image and the thumbnail, and then deletes the original image. When displaying the updated image in the UI, the application locates the images in blob storage using the information maintained in the expense item entity.

> *It's possible to implement background processes so that they run at a scheduled time, or so that they can run for a long time with the ability to restart part way through in the event of a failure, by using the features available in Windows Azure storage queues. For more information see "Windows Azure Queues Improved Leases, Progress Tracking, and Scheduling of Future Work."*

Using Windows Azure Storage Queues

When using a Windows Azure queue it is possible that a message could be read more than once. Different worker roles may try to read the same message, or a single worker role may attempt to read it again. For example, in the aExpense application a duplicate message would cause the worker role to try to resize the original image and to generate the thumbnail a second time, overwriting the saved compressed image and thumbnail. In addition to the unnecessary processing, this will fail if the previous process has already deleted the original uploaded image.

With Windows Azure queues, it is possible that a message could be read twice by the same worker role, or by two different worker roles.

To prevent this, message processing tasks must be *idempotent*. Either the task execution must not in any way affect the operation of the application or the integrity of the application's data; or the code must prevent the task being executed more than once.

If your message processing method is not inherently idempotent, there are several strategies that you can adopt to stop the message recipient processing a message multiple times:

- When you read a message from a queue, you can use the **visibility-Timeout** parameter to set how long the messages should be hidden from other readers (the default value is 30 seconds). This gives you time to make sure that you can process and delete the message from the queue before another client reads it. Getting a message from a queue does not automatically delete it from the queue. It is still possible for the **visibilityTimeout** period to expire before you delete the message; for example, if the method processing the message fails.

- Each message has a **DequeueCount** property that records how many times the message has been retrieved from the queue. However, if you rely on this property, and only process messages that have a **DequeueCount** value of **0**, your application must guard against the possibility that a message has been dequeued but not processed.

- You could also add a unique transaction identifier to the message and then save the identifier in the blob's metadata properties. If, when you retrieve a message from a queue, the unique identifier in the message matches the unique identifier in the blob's metadata, you know that the message has already been processed once.

Handling Transient Faults when Accessing Windows Azure Storage

Access to Window Azure storage is by using a URL that specifies the storage account and the path to the data items. This request will pass over the datacenter network if the storage account you use is in the same datacenter as the code making the request, which should typically be the case to achieve optimum performance. However, it is still possible that requests may fail to connect immediately due to transient network loads. This is especially the case if you need to access the storage account from outside that datacenter.

The Windows Azure storage client API includes support for custom retry policies. However, using the Transient Fault Handling Application Block enables you to take advantage of the customization offered by the block, use configuration to define your retry policies, and adopt a standard approach to all the retry logic in your application.

In Chapter 4, "Moving to Windows Azure SQL Database," you saw how Adatum uses the Enterprise Library Transient Fault Handling Block to provide a reliable retry mechanism to connecting to a database. The application block can also be used to connect to Windows Azure storage (blobs, queues, and tables). Adatum uses the application block in the UI part of the aExpense application to connect to the blob and queue when writing the uploaded image and posting a message to the worker role. It also uses the application block within the worker role for reading from and writing to blob storage.

Controlling Access to Receipt Images

The application allows users to view only images of receipts that they previously uploaded, or images that belong to business expense submissions that they can approve. The application keeps other images hidden to protect users' privacy. Adatum evaluated a several approaches for achieving this.

In Windows Azure, all storage mechanisms can be configured to allow data to be read from anywhere by using anonymous requests, which makes the model shown in Figure 3 very easy to implement:

> We evaluated three alternative approaches to making business expense receipt images browsable before deciding on shared access signatures.

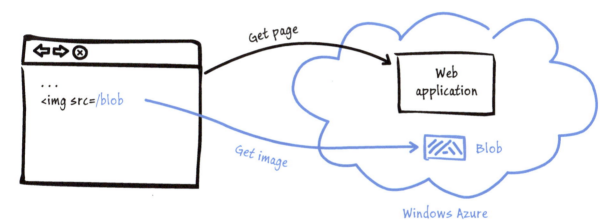

FIGURE 3
Directly addressable storage

In this scenario, you can access blob content directly through a URL such as **https://<application>.blob.core.windows.net/<containername>/<blobname>**. In the aExpense application, you could save the URLs as part of the expense entities in Windows Azure table storage. The advantages of this approach are its simplicity, the fact that data is cacheable, that it offloads work from the web server, and that it could leverage the Content Delivery Network (CDN) infrastructure in Windows Azure.

However, the disadvantage of using directly addressable blob storage is the lack of any security. Therefore, this is not an appropriate approach for the aExpense application because it would make it easy for someone to guess the address of a stored image, although this approach would work well for data that you did want to make publicly available such as logos, branding, or downloadable brochures.

> CDN enables you to cache blob data at strategically placed locations for delivering the content to users with the maximum available bandwidth.

Using deliberately obscure and complex URLs is a possible option, but this approach offers only weak protection and is not recommended.

The second approach considered by Adatum for accessing receipt images in blob storage was to route the request through the web site in much the same way that a "traditional" tiered application routes requests for data through the middle tier. Figure 4 shows this model.

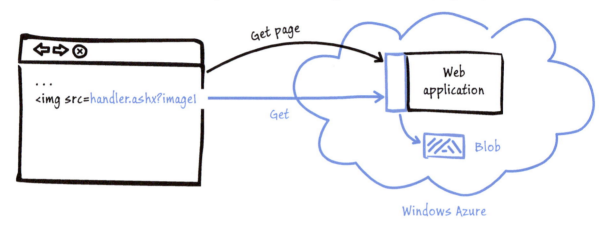

FIGURE 4
Routing image requests through the web server

In this scenario there is no public access to the blob container, and the web application holds the access keys for blob storage. This is how aExpense writes the image data to blob storage. A possible implementation of this scenario would be to use an HTTP handler to intercept image requests, check the access rules, and return the data. Although this approach would enable Adatum to control access to the images, it would add to the complexity of the web application and increase the workload of the web server. In addition, you couldn't use this approach if you wanted to use the Windows Azure CDN feature.

The approach that Adatum decided on for the aExpense application was to use the Windows Azure Shared Access Signature (SAS) feature. SAS enables you to control access to individual blobs by generating blob access URLs that are valid for defined period of time (though the valid time period can be unlimited if required). In the aExpense application, the web role generates these special URLs and embeds them in the page served to users. These special URLs then allow direct access to the blob for a limited period of time.

> SAS works well from a security perspective; the URL is only valid for a limited period of time, and would be difficult to guess. However, beware of exposing the blob data over the CDN because it is cached there and so the SAS lifetimes may be extended due to this caching.

There is some additional work for the web server because it must generate the SAS URLs, but Windows Azure blob storage handles most of the work. The approach is reasonably secure because the SAS URLs, in addition to having a limited lifetime, also contain a uniquely generated signature, which makes it very difficult for anyone to guess a URL before it expires.

INSIDE THE IMPLEMENTATION

Now is a good time to walk through these changes in more detail. As you go through this section, you may want to download the Visual Studio solution from *http://wag.codeplex.com/*. This solution (in the **Azure-WorkerRole** folder) contains the implementation of aExpense after the changes in this phase are made. If you are not interested in the mechanics, you should skip to the next section.

Uploading and Saving Images

In the aExpense application, the web role is responsible for uploading the image from the user's workstation and saving the initial, uncompressed version of the image. The following code in the **SaveExpense** method in the **ExpenseRepository** class (located in the **DataAccessLayer** folder of the **aExpense.Shared** project) calls the **AddReceipt** method of the **ExpenseReceiptStorage** class for each expense item that the user submits.

C#
```
this.receiptStorage.AddReceipt(expenseItem.Id.ToString(),
        expenseItem.Receipt, string.Empty);
```

The following code from the **ExpenseReceiptStorage** class shows how the **AddReceipt** method saves the original, uncompressed image to blob storage. Notice how it uses the Transient Fault Handling Block to retry all storage access operations.

C#
```
public string AddReceipt(string receiptId, byte[] receipt,
    string contentType)
{
  CloudBlob blob = this.storageRetryPolicy.ExecuteAction(
            () => this.container.GetBlobReference(receiptId));
  blob.Properties.ContentType = contentType;
  this.storageRetryPolicy.ExecuteAction(
            () => blob.UploadByteArray(receipt));
  return blob.Uri.ToString();
}
```

The Hands-on Labs that accompany this guide provide a step-by-step walkthrough of parts of the implementation tasks Adatum carried out on the aExpense application at this stage of the migration process.

*The retry policy used in the **AddReceipt** method is initialized in the class constructor by calling the **GetDefaultAzureStorage-RetryPolicy** method of the Transient Fault Handling Block's static **RetryPolicyFactory** class.*

Abstracting the Worker Role

Figure 5 summarizes the common pattern for the interaction between web roles and worker roles in Windows Azure.

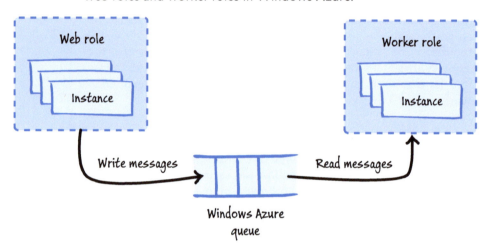

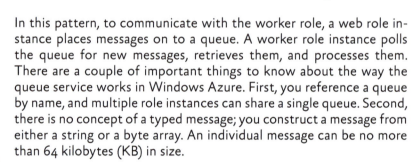

FIGURE 5
Web-to-worker role communication with a Windows Azure queue

In this pattern, to communicate with the worker role, a web role instance places messages on to a queue. A worker role instance polls the queue for new messages, retrieves them, and processes them. There are a couple of important things to know about the way the queue service works in Windows Azure. First, you reference a queue by name, and multiple role instances can share a single queue. Second, there is no concept of a typed message; you construct a message from either a string or a byte array. An individual message can be no more than 64 kilobytes (KB) in size.

> *If the size of your messages could be close to the maximum, be aware that Windows Azure converts all messages to Base64 before it adds them to the queue.*

Using a Windows Azure storage queue is a very common pattern for communicating between the web role and the worker role.

In addition, Windows Azure implements an "at-least-once" delivery mechanism; thus, it does not guarantee to deliver messages on a first-in, first-out basis, or to deliver only a single copy of a message, so your application should handle these possibilities.

In the current phase of the migration of aExpense to Windows Azure, the worker role only performs a single task. However, Adatum expects the worker role to take on additional responsibilities in later phases. Therefore, the team at Adatum developed some simple "plumbing" code to hide some of the complexities of Windows Azure worker roles and Windows Azure queues and to make them easier to work with in the future. Figure 6 is a high-level description of these abstractions and shows where to plug in your custom worker role functionality.

Windows Azure does not guarantee to deliver messages on a first-in, first-out basis, or to deliver only a single copy of a message.

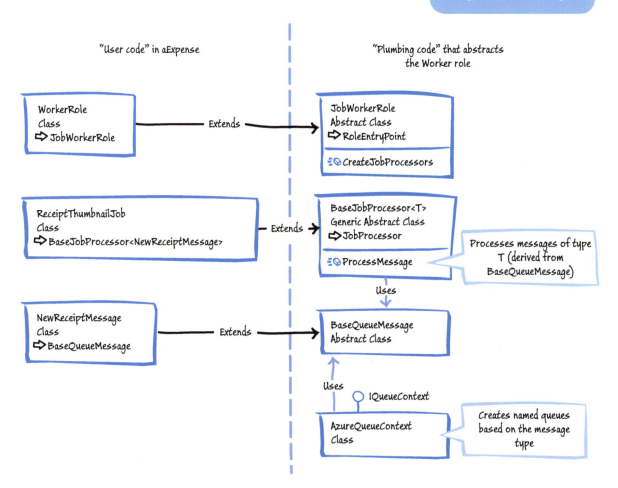

FIGURE 6
Relationship of user code to plumbing code

The "user code" classes are the ones that you will implement for each worker role and job type. The "plumbing code" classes are the re-usable elements. The plumbing code classes are packaged in the **AExpense. Jobs**, **AExpense.Queues**, and **AExpense.QueueMessages** namespaces (the queue and message classes are in the **aExpense.Shared** project). The following sections first discuss the user code and then the plumbing code.

User Code in the aExpense Application

The code you'll see described in this section implements the job that will compress images and generate thumbnails in the worker role for the aExpense application. What you should see in this section is how easy it is to define a new job type for the worker role to perform. This code uses the plumbing code that the following section describes in detail.

The following code in the **WorkerRole** class shows how the application initializes the worker role using the plumbing code. You create a new class that derives from the **JobWorkerRole** and override the **CreateJobProcessors** method. In this method, you instantiate your job processing objects that implement the **IJobProcessor** interface. As you can see, this approach makes it easy to plug in any additional job types that implement the **IJobProcessor** interface.

```csharp
public class WorkerRole : JobWorkerRole
{
  protected override IEnumerable<IJobProcessor>
      CreateJobProcessors()
  {
    return new IJobProcessor[] { new ReceiptThumbnailJob() };
  }
}
```

The constructor for the **ReceiptThumbnailJob** class specifies the interval the worker role uses to poll the queue. It instantiates an **AzureQueueContext** object, an **ExpenseReceiptStorage** object, and an **ExpenseRepository** object for the worker role to use, as shown here.

```csharp
C#
public ReceiptThumbnailJob()
    : base(2000, new AzureQueueContext())
{
  this.receiptStorage = new ExpenseReceiptStorage();
  this.expenseRepository = new ExpenseRepository();
}
```

The plumbing code passes a **NewReceiptMessage** object that contains the details of the image to process to the **ProcessMessage** method. This method then compresses the image referenced by the message and generates a thumbnail. The following code shows the **ProcessMessage** method in the **ReceiptThumbnailJob** class.

```csharp
C#
public override bool ProcessMessage(NewReceiptMessage message)
{
  …
}
```

Worker roles must poll the queue for new messages.

In the aExpense application, to send a message containing details of a new receipt image to the worker role, the web role creates a **NewReceiptMessage** object and calls the **AddMessage** method of the **AzureQueueContext** class. The following code shows the definition of the **NewReceiptMessage** class.

```csharp
C#
[DataContract]
public class NewReceiptMessage : BaseQueueMessage
{
    [DataMember]
    public string ExpenseItemId { get; set; }
}
```

*It's important to use the **DataContract** and **DataMember** attributes in your message class because the **AzureQueue-Context** class serializes message instances to the JSON format.*

The last few lines of code in the **SaveExpense** method of the **ExpenseRepository** class show how the web role in aExpense posts a message onto the queue.

```csharp
public void SaveExpense(Expense expense)
{
  using (var db = new ExpensesDataContext(
                     this.expenseDatabaseConnectionString))
  {
    var entity = expense.ToEntity();
    db.Expenses.InsertOnSubmit(entity);

    foreach (var detail in expense.Details)
    {
      var detailEntity = detail.ToEntity(expense);
      db.ExpenseDetails.InsertOnSubmit(detailEntity);
      if (detail.Receipt != null && detail.Receipt.Length > 0)
      {
        this.receiptStorage.AddReceipt(detail.Id.ToString(),
                          detail.Receipt, string.Empty);
      }
    }
    this.sqlCommandRetryPolicy.ExecuteAction(
          () => db.SubmitChanges());

    var queue = new AzureQueueContext(this.account);
    expense.Details.ToList().ForEach(i => queue.AddMessage(
          new NewReceiptMessage
                { ExpenseItemId = i.Id.ToString() })); }
}
```

Notice that the method does not place any messages on the queue to notify the worker role that new images are ready for processing until after all the records have been saved. This way, there is no chance that the worker role will try to process an image before the associated record has been saved and fail because it can't find the record (remember that the worker role must update the URLs of the image and thumbnail in the detail record).

The Plumbing Code Classes

Adatum developed the worker role abstractions to simplify the way that it sends messages from a web role to a worker role, and to simplify the way that it codes a worker role. The idea is that code can put a typed message onto a queue, and when the worker role retrieves the message, it routes it to the correct job processor for that message type. The previous section of this chapter described the job processor in the aExpense application that processes scanned receipt images using these abstractions. This section describes the abstractions in more detail.

The plumbing code classes simplify the way that you send messages from a web role to a worker role and the way that you implement a worker role.

The following core elements make up these plumbing code classes:

- A wrapper for the standard Windows Azure worker role's **RoleEntryPoint** class named **JobWorkerRole** that abstracts the worker role's life cycle and threading behavior.
- A customizable processor class named **JobProcessor** that enables users of the plumbing code classes to define their own job types for the worker role.
- A wrapper for the standard Windows Azure **CloudQueue** class named **AzureQueueContext** that implements typed messages to enable message routing within the **JobWorkerRole** class.

Figure 7 summarizes how the plumbing code classes handle messages derived from the **BaseQueueMessage** class.

Adatum expects to implement additional background processes, so it makes sense to create this plumbing code.

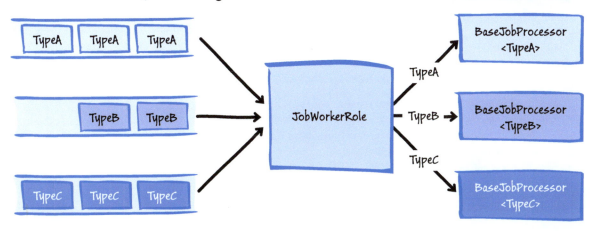

FIGURE 7
Worker role plumbing code elements

The message types that the plumbing code classes handle (such as the **NewReceiptMessage** type in aExpense) are derived from the **BaseQueueMessage** class shown in the following code example.

```csharp
C#
[DataContract]
public abstract class BaseQueueMessage
{
  private object context;

  public object GetContext()
  {
    return this.context;
  }

  public void SetContext(object value)
  {
    this.context = value;
  }
}
```

The plumbing code classes use the **AzureQueueContext** class to deliver messages to the worker role. The **AzureQueueContext** class creates named queues based on the message types, one queue for each message type that was registered with it. The following code shows the **AddMessage** method in the **AzureQueueContext** class that you use to add a new message to a queue, and the **Resolve-QueueName** method that figures out the name of the queue to use.

```csharp
C#
public void AddMessage(BaseQueueMessage message)
{
  var queueName = ResolveQueueName(message.GetType());

  var json = Serialize(message.GetType(), message);
  var cloudQueue = this.storageRetryPolicy.ExecuteAction(
      () => this.queue.GetQueueReference(queueName));
  this.storageRetryPolicy.ExecuteAction(
      () => cloudQueue.AddMessage(new CloudQueueMessage(json)));
}

public static string ResolveQueueName(MemberInfo messageType)
{
  return messageType.Name.ToLowerInvariant();
}
```

Notice that the plumbing code serializes messages to the JSON format, which typically produces smaller message sizes than an XML encoding (but possibly larger than a binary encoding).

The **AzureQueueContext** class uses a Windows Azure queue, and it's important that this queue has been created before performing any operations against it. The public **Purge** method of the **AzureQueueContext** class calls a private method named **EnsureQueueExists** that itself calls the **CreateIfNotExist** method of the Windows Azure **CloudQueue** class. Calling the **CreateIfNotExist** method counts as a storage transaction and will add to your application's running costs. To minimize the number of calls made to the **CreateIfNotExist** method the **AzureQueueContext** class maintains a list of queue names that it has already checked, as shown here.

```csharp
C#
private readonly ICollection<string> ensuredQueues;
…
private void EnsureQueueExists(string queueName)
{
  if (!this.ensuredQueues.Contains(queueName))
  {
    this.queue.GetQueueReference(queueName).CreateIfNotExist();
    this.ensuredQueues.Add(queueName);
  }
}
```

In Chapter 7, "Moving to Windows Azure Table Storage," the way that the application calls the **CreateIfNotExist** is revaluated following the results from performance testing.

The plumbing code classes deliver messages to job processor components, where a job processor handles a specific message type. The plumbing code classes include an interface named **IJobProcessor** that defines two void methods named **Run** and **Stop** for starting and stopping a processor. The abstract **BaseJobProcessor** and **JobProcessor** classes implement this interface. In the aExpense application, the **ReceiptThumbnailJob** class that you've already seen extends the **BaseJobProcessor** class.

If you are concerned about the running costs of the application, you should be aware of which calls in your code are chargeable! If you have high volumes of messages, you should check how frequently your application calls methods such as **CreateIfNotExist**.

The following code example shows how the **JobProcessor** class implements the **IJobProcessor** interface.

```csharp
C#
private bool keepRunning;

public void Stop()
{
  this.keepRunning = false;
}

public void Run()
{
  this.keepRunning = true;
  while (this.keepRunning)
  {
    Thread.Sleep(this.SleepInterval);
    this.RunCore();
  }
}

protected abstract void RunCore();
```

The following code example shows how the **BaseJobProcessor** class provides an implementation of the **RunCore** method.

```csharp
C#
protected bool RetrieveMultiple { get; set; }
protected int RetrieveMultipleMaxMessages { get; set; }

protected override void RunCore()
{
  if (this.RetrieveMultiple)
  {
    var messages = this.Queue.GetMultipleMessages<T>
                  (this.RetrieveMultipleMaxMessages);
    if (messages != null)
    {
      foreach (var message in messages)
      {
        this.ProcessMessageCore(message);
      }
    }
```

We don't use the Transient Fault Handling Block when reading messages from a queue because the worker role regularly polls the queue to receive messages, and so a failure during one polling cycle will leave the message in the queue ready to be received in the next one.

```
    else
    {
      this.OnEmptyQueue();
    }
  }
  else
  {
    var message = this.Queue.GetMessage<T>();
    if (message != null)
    {
      this.ProcessMessageCore(message);
    }
    else
    {
      this.OnEmptyQueue();
    }
  }
}
```

As you can see, the **RunCore** method can retrieve multiple messages from the queue in one go. The advantage of this approach is that one call to the **GetMessages** method of the Windows Azure **Cloud-Queue** class only counts as a single storage transaction, regardless of the number of messages it retrieves. The code example also shows how the **BaseJobProcessor** class calls the generic **GetMessage** and **GetMultipleMessages** methods of the **AzureQueueContext** class, specifying the message type by using a generic type parameter.

The following code example shows how the **BaseJobProcessor** constructor assigns the job's polling interval and the **AzureQueue-Context** reference.

C#
```
protected BaseJobProcessor(int sleepInterval,
        IQueueContext queue) : base(sleepInterval)
{
  if (queue == null)
  {
    throw new ArgumentNullException("queue");
  }
  this.Queue = queue;
}
```

It's cheaper and more efficient to retrieve multiple messages in one go if you can. However, these benefits must be balanced against the fact that it will take longer to process multiple messages; this risks the messages becoming visible to other queue readers before you delete them.

The remaining significant methods in the **BaseJobProcessor** class are the **ProcessMessageCore** and the abstract **ProcessMessage** methods shown below.

```csharp
C#
protected int MessagesProcessed { get; set; }

private void ProcessMessageCore(T message)
{
  var processed = this.ProcessMessage(message);
  if (processed)
  {
    this.Queue.DeleteMessage(message);
    this.MessagesProcessed++;
  }
}

public abstract bool ProcessMessage(T message);
```

The **RunCore** method in the **BaseJobProcessor** class invokes the **ProcessMessageCore** method shown above when it finds new messages to process. The **ProcessMessageCore** method then calls the "user-supplied" implementation of the **ProcessMessage** method before it deletes the message from the queue. In the aExpense application, this implementation is in the **ReceiptThumbnailJob** class.

The final component of the plumbing code is the abstract **JobWorkerRole** class that wraps the standard Windows Azure **RoleEntryPoint** class for the worker role. The following code example shows the **Run** method in this class.

```csharp
C#
protected IEnumerable<IJobProcessor> Processors { get; set; }

protected abstract IEnumerable<IJobProcessor>
                            CreateJobProcessors();

public override void Run()
{
  this.keepRunning = true;
  this.Processors = this.CreateJobProcessors();
  this.Tasks = new List<Task>();

  foreach (var processor in this.Processors)
  {
    var t = Task.Factory.StartNew(processor.Run);
    this.Tasks.Add(t);
  }
```

```
// Control and restart a faulted job
while (this.keepRunning)
{
  for (int i = 0; i < this.Tasks.Count; i++)
  {
    var task = this.Tasks[i];
    if (task.IsFaulted)
    {
      // Observe unhandled exception
      if (task.Exception != null)
      {
        Trace.TraceError("Job threw an exception: "
                + task.Exception.InnerException.Message);
      }
      else
      {
        Trace.TraceError("Job Failed no exception thrown.");
      }
      var jobToRestart = this.Processors.ElementAt(i);
      this.Tasks[i] = Task.Factory.StartNew(jobToRestart.Run);
    }
  }
  Thread.Sleep(TimeSpan.FromSeconds(30));
}
}
```

The **Run** method invokes the abstract **CreateJobProcessors** method that is implemented in user code. In the aExpense application, you can find the implementation of the **CreateJobProcessors** method in the **WorkerRole** class. The **Run** method then creates a new task for each job processor, monitors the state so that it can restart any that fail, and then waits for all the tasks to finish.

You need to keep the tasks within the worker role alive!

Processing the Images

The following code example shows how the aExpense application implements the image processing functionality in the **ProcessMessage** method in the **ReceiptThumbnailJob** class.

```csharp
C#
public override bool ProcessMessage(NewReceiptMessage message)
{
  var expenseItemId = message.ExpenseItemId;
  var imageName = expenseItemId + ".jpg";

  byte[] originalReceipt
          = this.receiptStorage.GetReceipt(expenseItemId);

  if (originalReceipt != null && originalReceipt.Length > 0)
  {
    var thumb = ResizeImage(originalReceipt, ThumbnailSize);
    var thumbUri = this.receiptStorage.AddReceipt(
                       Path.Combine("thumbnails", imageName),
                                    thumb, "image/jpeg");

    var receipt = ResizeImage(originalReceipt, PhotoSize);
    var receiptUri = this.receiptStorage.AddReceipt(
                         imageName, receipt, "image/jpeg");

    this.expenseRepository.UpdateExpenseItemImages(
                       expenseItemId, receiptUri, thumbUri);
    this.receiptStorage.DeleteReceipt(expenseItemId);

    return true;
  }

  return false;
}
```

Although we limit users to uploading images that are less than 1,024 KB in size, we decided not to store the original images in order to save space. We found that resizing to a standard size provided acceptable quality.

This method retrieves the image name from the message sent to the worker role and creates two new versions of the image: one a thumbnail, and one a fixed standard size. It then deletes the original image. The method can process images in any standard format, but it always saves images as JPEGs.

> The **ProcessMessage** method should be idempotent, so there are no unwanted side effects if a message is delivered multiple times. The **ProcessMessage** method should also contain some logic to handle "poison" messages that cannot be processed for any reason.

Making the Images Available Using Shared Access Signatures

To make images of receipts viewable in the UI, the team at Adatum used Shared Access Signatures (SAS) to generate short-lived, secure URLs to address the images in blob storage. This approach avoids having to give public access to the blob container, and minimizes the amount of work that the web server has to perform because the client can access the image directly from blob storage.

The following code example shows how the application generates the SAS URLs in the **GetExpenseByID** method in the **Expense-Repository** class by appending the SAS to the blob URL. The aExpense application uses an HTTPS endpoint, so the blob reference and signature elements of the blob's URL are protected by SSL from "man-in-the-middle" attacks.

The aExpense application uses Shared Access Signatures (SAS) to provide limited access to blobs in private containers. As well as controlling access to blobs at container level you can configure SAS for individual blobs, queues, tables, and partition or row key ranges in a table.

```C#
CloudBlob receiptBlob = this.storageRetryPolicy.ExecuteAction(
  () => container.GetBlobReference(item.ReceiptUrl.ToString()));
item.ReceiptUrl = new Uri(item.ReceiptUrl.AbsoluteUri +
    receiptBlob.GetSharedAccessSignature(policy));
```

The **GetSharedAccessSignature** method takes a **SharedAccessPolicy** object as a parameter. This policy specifies the access permissions and the lifetime of the generated URL. The following code shows the policy that the aExpense application uses to grant read permission for an image. The application generates a new SAS whenever a user tries to access an expense submission.

The Shared Access Signature is appended to the standard blob URL. Using SSL ensures that all URL data except for the hostname is encrypted.

```C#
private readonly TimeSpan sharedSignatureValiditySpan;

var policy = new SharedAccessPolicy
{
  Permissions = SharedAccessPermissions.Read,
  SharedAccessStartTime = DateTime.UtcNow.AddMinutes(-5),
  SharedAccessExpiryTime = DateTime.UtcNow +
                           this.sharedSignatureValiditySpan
};
```

The value of the **Timespan** variable named **sharedSignatureValiditySpan** is set in the constructor of the the **ExpenseRepository** class to the same value as the default ASP.NET session timeout. Notice that the code specifies a value five minutes before the current time on the server for the **SharedAccessStart-Time** property of the **SharedAccessPolicy** object. This is done to prevent any clock skew between the client and the server from preventing access if the user tries to access the blob immediately.

> *The request will succeed as long as the **Get** request for a blob starts after the start time and before the expiry time, even if the response streaming continues past the expiry time. In the aExpense application, the user's browser is the client that accesses the data stored in the blob. However, if your client application is using the **StorageClient** library to read the blob and chooses to retry on a **Get** failure as a part of the logical **Get** operation, which is the default **StorageClient** library policy, any retry request made after the expiration time will fail. The same will apply if you are using the Transient Fault Handling Application Block to handle retries. If you decide to have a short validity period for the URL, make sure that you issue a single **Get** request for the entire blob and use a custom retry policy so that if you retry the request, you get a new SAS for the URL.*

MORE INFORMATION

All links in this book are accessible from the book's online bibliography available at: *http://msdn.microsoft.com/en-us/library/ff803373.aspx.*

MSDN is a good starting point to learn about Windows Azure Blob Storage. Start at *"How to use the Windows Azure Blob Storage Service in .NET."*

To find out more about controlling access to Azure storage, including shared access signatures, look at *"Managing Access to Containers, Blobs, Tables, and Queues."*

You can find a summary of the Windows Azure service architecture at *"Overview of Creating a Hosted Service for Windows Azure."*

All links in this book are accessible from the book's online bibliography available at: *http://msdn.microsoft.com/en-us/library/ff803373.aspx.*

6

Evaluating Cloud Hosting Costs

This chapter presents a basic cost model for running the aExpense application in the cloud. It makes some assumptions about the usage of the application and uses the current pricing information for Windows Azure services to estimate annual operational costs.

THE PREMISE

The aExpense application is a typical business application. Adatum selected this as a pilot cloud migration project because the application has features that are common to many of Adatum's other business applications, and Adatum hopes that any lessons learned from the project can be applied elsewhere.

The original on-premises version of the aExpense application is deployed in Adatum's data center, with components installed across several different servers. The web application is hosted on a Windows Server computer that it shares with another application. aExpense also shares a SQL Server database installation with several other applications, but has its own dedicated drive array for storing scanned expense receipts.

The current cloud-based deployment of aExpense, using Cloud Services web and worker roles, is sized for average use, not peak use, so the application can be slow and unresponsive during the busy two days at month-end when the majority of users submit their business expense claims.

GOALS AND REQUIREMENTS

It is difficult for Adatum to determine accurately how much it costs to run the original on-premises version of aExpense. The application uses several different servers, shares an installation of SQL Server with several other business applications, and is backed up as part of the overall backup strategy in the data center.

It is very difficult to estimate the operational costs of an existing on-premises application.

Although Adatum cannot determine the existing running costs of the application, Adatum wants to estimate how much it will cost to run in the cloud now that the developers have completed the migration steps described in the previous chapters of this guide. One of the specific goals of the pilot project is to discover how accurately it can predict running costs for cloud based applications.

A second goal is to estimate what cost savings might be possible by configuring the application in different ways, or by taking advantage of other Windows Azure services. Adatum will then be able to assign a cost to a particular configuration and level of service, which will make it much easier to perform a cost-benefit analysis on the application. A specific example of this in the aExpense application is to estimate how much it will cost to deploy additional instances to meet peak demand during the busy month-end period.

Overall, Adatum would like to see greater transparency in managing the costs of its suite of business applications.

DETAILED COSTING ESTIMATES

The first step Adatum took was to analyze what it will be billed every month for the cloud-based version of aExpense. Figure 1 shows the services that Microsoft will bill Adatum for each month for the aExpense application.

> You can manage the cost of a cloud-based application by modifying its behavior through configuration changes.

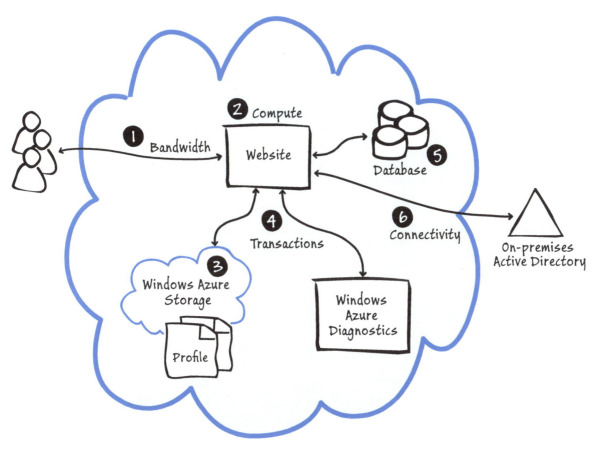

FIGURE 1
Billable services

The simple cost model in this chapter does not include any cost estimates for testing environments. Adatum should review its cost estimates when the aExpense application design is complete and when Adatum completes stress-testing the application. It also ignores the cost of connectivity to on-premises Active Directory or a database through Windows Azure Connect or Virtual Networks as this is only applicable to some of the original IaaS scenarios.

The following table summarizes the current rates in U.S. dollars for these services. The prices listed here are accurate for the U.S. market as of July 2012. However, for up-to-date pricing information see the Windows Azure *Pricing Details*. You can find the pricing for other regions at the same address.

Service	Description	Cost
1. In/Out Bandwidth	This is the web traffic between the user's browser and the aExpense site.	Inbound: Free Outbound (North America and Europe): $0.12 per GB
2. Compute	Virtual machines, for the time each one is running.	Small size virtual machine: $0.115 per hour Medium size virtual machine: $0.23 per hour
	Cloud Services roles, for the time each role is running.	Small size role: $0.12 per hour Medium size role: $0.24 per hour
3. Windows Azure Storage	In aExpense this will be used to store scanned receipt images. Later, it will also store profile data when Adatum removes the requirement for a relational database.	Up to 1 TB with geo-replication: $0.125 per GB Up to 1 TB without geo-replication: $0.09 per GB
4. Transactions	Each interaction with the storage system is billed.	$0.01 per 100,000 transactions
5. Database	SQL Server hosted in a VM	Small or medium size VM: $0.55 per hour
	Windows Azure SQL Database, cost per month.	Up to 100 MB: $4.995 Up to 1 GB: $9.99 Up to 10 GB: First GB $9.99, each additional GB $3.996 Up to 50 GB: First 10 GB $45.954, each additional GB $1.998 Up to 150 GB: First 50 GB $125.874, each additional GB $0.999
6. Connectivity	Virtual Networks and Connect	$0.05 per hour per connection

After you have estimated your usage of the billable services by your application, you can use the Windows Azure *pricing calculator* to quickly estimate your monthly costs.

Bandwidth Cost Estimate for aExpense

The aExpense application is not bandwidth intensive. Assuming that all scanned receipt images will be transferred back and forth to the application twice, and taking into account the web traffic for the application, Adatum estimated that 9.5 GB of data would move each way every month.

Data transfer	GB/month	$/GB/month	Total/month
Inbound	9.5 GB	Currently free	$ 0.00
Outbound	9.5 GB	$ 0.12	$ 1.14
		Total/year	$ 13.68

> The Hands-on Labs that are available for this guide include an exercise that demonstrates how Adatum estimated the bandwidth usage and other runtime parameters for the aExpense application.

Compute Estimate for aExpense

Adatum's assumption here is that the application will run 24 hours a day, 365 days a year. The current version of the application uses a single instance of the Cloud Services web role and worker role.

Hours (one year)	$/hour	Number of role instances	Total/year
8760	$ 0.12	2	$ 2,102.40

Receipt Image Storage Estimate for aExpense

The aExpense application stores uploaded receipt images in Windows Azure blob storage. Based on an analysis of existing usage, on average 65 percent of 15,000 Adatum employees submit ten business expense items per month. Each scanned receipt averages 15 KB in size, and to meet regulatory requirements, the application must store seven years of history. This gives an estimated storage requirement for the application of 120 GB.

Storage and transactions	Amount	Cost per month	Total/month
Total GB stored	120 GB	$ 0.125/ GB	$ 15.14
Storage transactions per month	700,000	$ 0.01/100 K	$ 0.07
		Total/year	$ 182.52

Windows Azure SQL Database Storage Requirements Estimate

The aExpense application stores expense data (other than the receipt images) in a Windows Azure SQL Database. Adatum estimates that each business expense record in the database will require 2 KB of storage. So based on the analysis of existing usage (on average 65 percent of 15,000 Adatum employees submit ten business expense items per month) and the requirement to store data for seven years, this gives an estimated storage requirement of 16 GB. However, the actual measured database usage is likely to be greater than this due to the nature of the way that a database stores the data until it is compacted, and so Adatum will base the estimate on a 20 GB database.

SQL storage size	$/month	Total/year
20 GB	$ 65.93	$ 791.16

Total Cost Approximation

This means that the costs as an approximate proportion of the total cost of running the application (a total of $ 3,089.76 per year) will be as follows:

- Compute (web and worker roles): $ 2,102.40 (~ 68 %)
- Windows Azure SQL Database: $ 791.16 (~ 26 %)
- Windows Azure storage: $ 182.52 (~ 6 %)
- Bandwidth: $ 13.68 (~ 0.4 %)

VARIATIONS

Having established the approximate costs of running the aExpense application in its current form, Adatum wants to confirm that its choice of a PaaS approach was justified, and also consider some variations to discover the cost for better meeting peak demand and to see if additional cost savings were possible.

Costing the IaaS Hosting Approach

In the first step of the migration, Adatum hosted both the application and the SQL Server database in Windows Azure Virtual Machines. To accommodate the requirements of SQL Server, with a view to using it with other applications in the future, Adatum chose to use a medium sized virtual machine for the database, and a small sized virtual machine for the application.

Adatum also chose to use the Standard edition of SQL Server rather than the Web edition in order to accommodate future requirements. The virtual machine that hosts SQL Server also needs a data disk to store the SQL Server database. The estimated costs of this configuration are shown in the following table.

Service	$/month	Total/year
Small virtual machine	$ 83.90	$ 1,006.80
Medium virtual machine	$ 167.81	$ 2,013.72
Data Disk (20 GB)	$ 2.50	$ 30.00
SQL Server	$ 401.28	$ 4,815.36
Bandwidth usage	$ 13.68	$ 164.16
Virtual network	$ 36.48	$ 437.76
Total		**$ 8,467.80**

From this it's clear that the PaaS approach using Cloud Services and Windows Azure SQL Database is considerably less expensive than the original migration step that used the IaaS approach with virtual machines. However, Adatum must consider that the IaaS approach required only very minimal changes to the application code, and that the use of a hosted SQL Server is not directly equivalent to using Windows Azure SQL Database. For example, if Adatum deploys additional applications in the future they can share the hosted SQL Server without incurring additional cost, whereas additional costs will be incurred when other applications that use Windows Azure SQL Database are deployed.

It's also possible for Adatum to install SQL Server on the virtual machine using a licensed copy they own instead of paying to rent SQL Server, which could considerably reduce the overall costs; but Adatum must also consider the cost of maintaining and upgrading the operating systems and database software for the IaaS approach.

However, overall, the saving of almost $ 5,000.00 per year justifies the decision Adatum made to move from IaaS to PaaS for the aExpense application, even when considering the development effort required to refactor the application and adapt it to run in Cloud Services web and worker roles. Adatum will review the situation when it decides to move other applications to the cloud.

Adatum chose to use the preinstalled SQL Server Standard edition that is available in the Windows Azure Management Portal with a view to sharing the database with other applications in the future. However, if it chose to use the Web edition, the monthly cost for SQL Server falls from $401.28 to $32.83, and the annual cost falls to $ 393.98. The total annual cost of this IaaS approach now becomes $4,046.42.

Combined IaaS and PaaS Approach

If Adatum actually requires a virtual machine because the application demands some special operating system configuration, access to non-standard installed services, or cannot be refactored into web and worker roles, the data could be still be stored in Windows Azure SQL Database to remove the requirement for a hosted SQL Server. In this case the running costs per instance would be similar to that for the PaaS approach using Cloud Services.

Service	$/month	Total/year
Small virtual machine	$ 83.90	$ 1,006.80
Windows Azure SQL Database	$ 65.93	$ 791.16
Bandwidth usage	$ 13.68	$ 164.16
	Total/year	**$ 1,962.12**

This configuration is based on a single virtual machine, which would run the background tasks asynchronously within the application instead of using a separate worker role. Adatum could implement a virtual network in the cloud and load balance two virtual machine instances to provide additional capacity, in which case the overall cost would be almost the same as using a Cloud Services web and worker role.

Adatum could also use a virtual machine with a separate Cloud Services worker role to perform the background processing tasks, and communicate with the worker role from the virtual machine using Windows Azure storage queues. This configuration will also cost almost the same as when using Cloud Services web and worker roles.

The hourly compute costs for virtual machines, Cloud Services roles, and Windows Azure Web Sites Reserved instances (when all are generally available at the end of the discounted trial period) are almost the same, and so the decision on which to choose should be based on application requirements rather than focusing on just the compute cost.

Costing for Peak and Reduced Demand

One of the issues raised by users of the existing aExpense application is poor performance of the application during the two days at the end of the month when the application is most heavily used. To address this issue, Adatum then looked at the cost of doubling the compute capacity of the application for two days a month by adding an extra two web roles to handle the UI load.

Additional hours/month	Additional hours/year	$/hour	Role instances	$/year
48	576	$ 0.12	2	$ 138.24

This indicates that the additional cost to better meet peak demand is low, and yet it will provide a huge benefit for users. Adatum can use scripts executed by on-premises administrators to change the number of running instances, perhaps through a scheduled task, or implement an auto scaling solution such as the Enterprise Library Autoscaling Application Block.

> *The Autoscaling Application Block is part of Enterprise Library, developed by the p&p team at Microsoft. For more information see "The Autoscaling Application Block" and Chapter 6, "Maximizing Scalability, Availability, and Performance in the Orders Application," in the p&p guide "Building Hybrid Applications in the Cloud on Windows Azure."*

Adatum also examined the cost implications of running the application for only twelve hours each day for only six days each week, except at the month end when the majority of users access it. The following table shows the compute costs for the web and worker roles.

Compute	Number of role instances	Hours/day	Days/year	$/hour	$/year
Standard	2	12	313	$ 0.12	$ 901.44
Month End	2	12	24	$ 0.12	$ 69.12
				Total/year	$ 970.56

This is less than half of the compute cost of running the application 24 hours a day for 365 days per year, giving a saving of around $ 1,100 per year. Adatum could use the same auto scaling approach described earlier to achieve this pattern of availability.

Costing for Windows Azure Table Storage

Adatum is also interested in comparing the cost of storing the business expense data in Windows Azure table storage instead of in Windows Azure SQL Database. The previous calculations in this chapter show that the storage requirement for seven years of data (excluding receipt images, which are stored in Windows Azure blob storage) is around 16 GB. The following table also assumes that each new business expense item is accessed five times during the month.

Storage and transactions	Amount	Cost	Total/ Month
GB stored	16 GB	$ 0.125/ GB	$ 2.00
Storage transactions per month	220,000	$ 0.01/10 K	$ 0.02
		Total/year	$ 24.30

As you can see, this is a fraction of the cost of using Windows Azure SQL Database ($ 791.16 per year). The estimated total running cost for the year would be $ 2,322.90 using table storage, offering the possibility of reducing the overall running costs by almost a quarter.

Adapting the application to use Windows Azure table storage instead of a relational database will require development and testing effort. However, as long as table storage can provide the performance and scalability required, the saving makes this worthwhile for Adatum's scenario. In the following chapter you will see how Adatum explored adapting the aExpense application to use table storage, and then implemented the change.

MORE INFORMATION

All links in this book are accessible from the book's online bibliography available at: *http://msdn.microsoft.com/en-us/library/ff803373.aspx.*

Use the *Windows Azure Pricing* calculator to estimate runtime costs.

You can find information that will help you to understand your Windows Azure bill at *Pricing Details.*

For information about auto scaling Windows Azure application roles, see *"The Autoscaling Application Block"* and Chapter 6, *"Maximizing Scalability, Availability, and Performance in the Orders Application,"* in the p&p guide *"Building Hybrid Applications in the Cloud on Windows Azure."*

All links in this book are accessible from the book's online bibliography available at: *http://msdn.microsoft.com/en-us/library/ff803373.aspx.*

7 Moving to Windows Azure Table Storage

This chapter describes Adatum's final step in its migration process to the cloud for the aExpense application. It discusses the advantages of using Windows Azure storage instead of a relational database for expense items, the design of a suitable schema for storage, and how the developers at Adatum adapted the data access functions of the application to use Windows Azure storage instead of a relational database. The chapter also walks through the data export feature that Adatum added to the aExpense application, and some of the changes the developers at Adatum made following performance testing.

THE PREMISE

Adatum has now completed the migration of the aExpense application to the cloud, and added functionality that was missing during the initial migration so that users can upload and view scanned receipt images. However, as Adatum discovered when revisiting the costs of running the application in the cloud, there is one more opportunity to minimize these costs by switching to use Windows Azure storage for expense items instead of a relational database.

Adatum also wants to add a final piece of functionality in the application. The aExpense application must generate a file of data that summarizes the approved business expense submissions for a period. Adatum's on-premises payments system imports this data file and then makes the payments to Adatum employees.

In addition to implementing these changes to the aExpense application, Adatum also needs to perform final performance testing and tuning to ensure that the application provides an optimum user experience whilst minimizing its resource usage.

GOALS AND REQUIREMENTS

In this phase, Adatum has several specific goals. A simple cost analysis of the existing solution has revealed that Windows Azure SQL Database would account for about one quarter of the annual running costs of the application (see Chapter 6, "Evaluating Cloud Hosting Costs," for details of the cost calculations). Because the cost of using Windows Azure storage is less than using Windows Azure SQL Database, Adatum is keen to investigate whether it can use Windows Azure storage instead.

*You should evaluate whether
Windows Azure storage can
replace relational database
storage in your application.*

Adatum must evaluate whether the aExpense application can use Windows Azure storage. Data integrity is critical, so Adatum wants to use transactions when a user submits multiple business expense items as a part of an expense submission.

Also in this phase of the aExpense migration the project the team at Adatum will create the data export feature for integration with its on-premises systems. The on-premises version of aExpense uses a scheduled SQL Server Integration Services job to generate the output file and sets the status of an expense submission to "processing" after it is exported. The on-premises application also imports data from the payments processing system to update the status of the expense submissions after the payment processing system makes a payment. This import process is not included in the current phase of the migration project.

Figure 1 summarizes the export process in the original on-premises application.

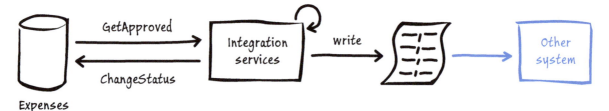

Expenses

FIGURE 1
The aExpense export process

The design of the export process for the cloud version of aExpense must meet a number of goals. First, the cost of the export process should be kept to a minimum while making sure that it does not have a negative impact on the performance of the application for users. The export process must also be robust and be able to recover from a failure without compromising the integrity of aExpense's data or the accuracy of the exported data.

The solution must also address the question of how to initiate the export by evaluating whether it should be a manually initiated operation or run on a specific schedule. If it is the latter, the team at Adatum must design a mechanism for initiating the task, such as using a Timer instance to execute it at regular intervals or by using a third party scheduler such as *Quartz*.

Approved business expense submissions could be anywhere in the table. We want to try to avoid the performance impact that would result from scanning the entire table.

The final requirement is to include a mechanism for transferring the data from the cloud-environment to the on-premises environment where the payment processing application can access it.

Adatum has also evaluated the results from performance testing the application, and needs to implement a number of changes based on those results. For example, the developers discovered that constantly checking for the existence of a queue or table before accessing it was causing unnecessary processing overhead, and decided that the application should initialize storage requirements only once during startup, removing the need to check for the existence on each call that reads or writes data.

The developers at Adatum also explored whether they should implement a paging mechanism, for displaying expense items, and how they could improve performance by fine tuning the configuration and the Windows Communication Foundation (WCF) Data Service code.

Adatum must automate the process of downloading the expense submission report data for input into the payments processing system.

Overview of the Solution

In this section you will see how the developers at Adatum considered the options available for meeting their goals in this stage of the migration process, and the decisions they made.

Why Use Windows Azure Table Storage?

As you saw in Chapter 5, "Executing Background Tasks," Adatum already uses Windows Azure storage blobs for storing the scanned receipt images and Windows Azure storage queues for transferring data between the web and worker roles. This functionality was added to the aExpense application during the migration step described in Chapter 5.

However, for storing data that is fundamentally relational in nature, such as the expense items currently stored in Windows Azure SQL Database, the most appropriate Windows Azure storage mechanism is tables. Windows Azure tables provide a non-relational table-structured storage mechanism. Tables are collections of entities that do not have an enforced schema, which means a single table can contain entities that have different sets of properties.

Even though the underlying approach is different from a relational database table, because each row is an entity that contains a collection of properties rather than a set of data rows containing columns of predefined data types, Windows Azure tables can provide an equivalent storage capability.

In Chapter 6, "Evaluating Cloud Hosting Costs," of this guide you discovered that Windows Azure table storage is less expensive per gigabyte stored than using Windows Azure SQL Database. For example, in Adatum's specific scenario, the running costs for the SQL Database are around $800.00 per year, which is 26% of the total cost. The calculated cost of the equivalent storage using Windows Azure table storage is only around $25.00 per year, which is less than 1% of the total running costs. Therefore, it makes sense financially to consider moving to table storage, as long as the development and testing costs are not excessive and performance can be maintained.

In addition to the cost advantage, Windows Azure tables also offer other useful capabilities. They can be used to store huge volumes of data (a single Windows Azure storage account can hold up to 100 TB of data), and can be accessed using a managed API or directly using REST queries. You can use Shared Access Signatures to control access to tables, partitions, and rows.

In some circumstances table storage can also provide better scalability. The data is also protected through automatic geo-replication across multiple datacenters unless you disable this function (for example, if legal restrictions prevent data from being co-located in other regions).

Profile Data

By moving the expenses data from Windows Azure SQL Database to Windows Azure table storage, Adatum will be able to remove the dependency of the aExpense application on a relational database. The justification for using table storage assumes that Adatum will no longer need to pay for a cloud hosted SQL Server or Windows Azure SQL Database.

However, when reviewing this decision, Adatum realized that the aExpense application still uses the ASP.NET profile provider, which stores user profile data in Windows Azure SQL Database. Therefore Adatum must find an alternative method for storing profile data.

Adatum uses Windows Azure Caching to store session data for users, but this is not suitable for storing profile data that must be persisted between user sessions. The developers at Adatum could write a custom profile provider that stores its data in Windows Azure storage. However, after investigation, they decided to use the *Windows Azure ASP.NET Providers sample*. This provider can be used to store membership, profile, roles, and session data in Windows Azure tables and blobs.

The Data Export Process

There are three elements of the export process to consider: how to initiate the process, how to generate the data, and how to download the data from the cloud.

Initiating the Export Process

The simplest option for initiating the data export is to have a web page that returns the data on request, but there are some potential disadvantages to this approach. First, it adds to the web server's load and potentially affects the other users of the system. In the case of aExpense, this will probably not be significant because the computational requirements for producing the report are low. Second, if the process that generates the data is complex and the data volumes are high, the web page must be able to handle timeouts. Again, for aExpense, it is unlikely that this will be a significant problem.

The most significant drawback to this solution in aExpense is that the current storage architecture for expense submission data is optimized for updating and retrieving individual expense submissions by using the user ID. The export process will need to access expense submission data by date and expense state. Unlike Windows Azure SQL Database where you can define multiple indexes on a table, Windows Azure table storage only has a single index on each table.

Figure 2 illustrates the second option for initiating the data export. Each task has a dedicated worker role, so the image compression and thumbnail generation would be handled by Task 1 in Worker 1, and the data export would be performed by Task 2 in Worker 2. This would also be simple to implement, but in the case of aExpense where the export process will run twice a month, it's not worth the overhead of having a separate role instance. If your task ran more frequently and if it was computationally intensive, you might consider an additional worker role.

Choosing the right Partition Key and Row Key for your tables is crucial for the performance of your application. Any process that needs a "table scan" across all your partitions will be slow.

You should try to use your compute nodes to the fullest. Remember, you pay for a deployed role instance whether or not it's doing any work. You can opt for a larger compute instance if you want to do more work in a single role instance.

FIGURE 2
Separate worker roles for each task

Figure 3 illustrates the third option where an additional task inside an existing worker role performs the data export process. This approach makes use of existing compute resources and makes sense if the tasks are not too computationally intensive. At the present time, the Windows Azure SDK does not include any task abstractions, so you need to either develop or find a framework to handle task-based processing for you. The team at Adatum will use the plumbing code classes described in Chapter 5, "Executing Background Tasks," to define the tasks in the aExpense application. Designing and building this type of framework is not very difficult, but you do need to include all your own error handling and scheduling logic.

Adatum already has some simple abstractions that enable them to run multiple tasks in a single worker role.

Windows Azure can only monitor at the level of a worker, and it tries to restart a failed worker if possible. If one of your task processing threads fails, it's up to you to handle the situation as described in Chapter 5.

Worker

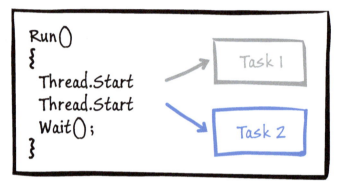

FIGURE 3
Multiple tasks in a single worker role

Generating the Export Data

The team at Adatum decided to split the expense report generation process into two steps. The first step "flattens" the data model and puts the data for export into a Windows Azure table. This table uses the expense submission's approval date as the partition key, the expense ID as the row key, and it stores the expense submission total. The second step reads this table and generates a Windows Azure blob that contains the data ready for export as a comma-separated values (CSV) file. Adatum implemented each of these two steps as a task by using the plumbing code described in Chapter 5, "Executing Background Tasks." Figure 4 illustrates how the task that adds data to the Windows Azure table works.

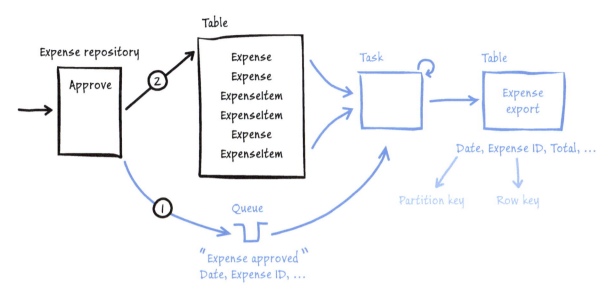

FIGURE 4
Generating the Expense Report table

First, a manager approves a business expense submission. This places a message that contains the expense submission's ID and approval date onto a queue (1), and updates the status of the submission in table storage (2). The task retrieves the message from the queue, calculates the total value of the expense submission from the expense detail items, and stores this as a single line in the Expense Export table. The task also updates the status of the expense submission to be "in process" before it deletes the message from the queue.

Exporting the Report Data

To export the data, Adatum considered two options. The first was to have a web page that enables a user to download the expense report data as a file. This page would query the expense report table by date and generate a CSV file that the payments processing system can import. Figure 5 illustrates this option.

Windows Azure table storage does not have all the features of a relational database, and so complex querying is more challenging. You may need multiple tables that present the same data in different ways based on the needs of the application. Table storage is cheap!

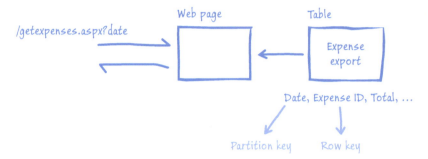

FIGURE 5
Downloading the expense report from a web page

The second option, shown in Figure 6, was to create another job in the worker process that runs on a schedule to generate the file in blob storage ready for download. Adatum will modify the on-premises payment processing system to download this file before importing it. Adatum selected this option because it enables them to schedule the job to run at a quiet time in order to avoid any impact on the performance of the application for users. The on-premises application can access the blob storage directly without involving either the Windows Azure web role or worker role.

> This approach makes it easier for us to automate the download and get the data in time for the payments processing run.

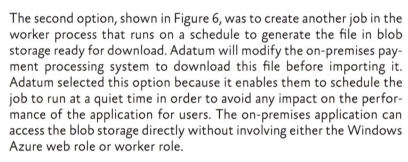

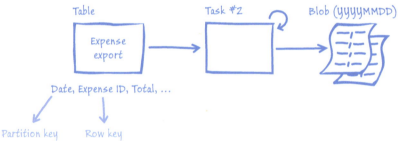

FIGURE 6
Generating the expense report in blob storage

Adatum had to modify slightly the worker role plumbing code to support this process. In the original version of the plumbing code, a message in a queue triggered a task to run, but the application now also requires the ability to schedule tasks.

> We had to modify our plumbing code classes slightly to accommodate scheduled tasks.

INSIDE THE IMPLEMENTATION

Now is a good time to walk through these changes in more detail. As you go through this section, you may want to download the Visual Studio solution from *http://wag.codeplex.com/*. This solution (in the **Azure-TableStorage** folder) contains the implementation of aExpense after the changes made in this phase. If you are not interested in the mechanics, you should skip to the next section.

The Hands-on Labs that accompany this guide provide a step-by-step walkthrough of parts of the implementation tasks Adatum carried out on the aExpense application at this stage of the migration process.

Storing Business Expense Data in Windows Azure Table Storage

Moving from Windows Azure SQL Database to Windows Azure table storage meant that the developers at Adatum had to re-implement the data access layer (DAL) in the application. The original version of aExpense used LINQ to SQL as the technology in the data access layer to communicate with Windows Azure SQL Database. The DAL converted the data that it retrieved using LINQ to SQL to a set of domain-model objects that it passed to the user interface (UI).

The new version of aExpense that uses Windows Azure table storage uses the managed Windows Azure storage client to interact with Windows Azure table storage. Because Windows Azure table storage uses a fundamentally different approach to storage, this was not simply a case of replacing LINQ to SQL with the .NET Client Library.

Keeping all your data access code in a data access layer restricts the scope of the code changes required if you need to change your storage solution (the code changes take place only in the DAL).

Use the Windows Azure storage client in the Windows Azure Managed Library to access Windows Azure table storage. Note that the Windows Azure table service only supports a subset of the functionality defined by the .NET Client Library for WCF Data Services. You can find more details in the topic *"Table Service Support for .NET Client Library Constructs."*

How Many Tables?

The most important thing to understand when transitioning to Windows Azure table storage is that the storage model is different from what you may be used to. In the relational world, the obvious data model for aExpense would have two tables, one for expense header entities and one for expense detail entities, with a foreign-key constraint to enforce data integrity. This reflects the schema that Adatum used in SQL Server and Windows Azure SQL Database in previous steps of the migration process.

However, the best data model to use is not so obvious with Windows Azure table storage for a number of reasons:

- You can store multiple entity types in a single table in Windows Azure.
- Entity Group Transactions are limited to a single partition in a single table (partitions are discussed in more detail later in this chapter).
- Windows Azure table storage is relatively cheap, so you shouldn't be so concerned about normalizing your data and eliminating data redundancy.

Adatum could have used two Windows Azure storage tables to store the expense header and expense detail entities. The advantage of this approach is simplicity because each table has its own, separate, schema. However, because transactions cannot span tables in Windows Azure storage, there is a possibility that orphaned detail records could be left if there was a failure before the aExpense application saved the header record.

For example, the developers would need to use two transactions to save an expense if Adatum had used two separate tables. The following code sample shows the outline of the **SaveExpense** method that would be required in the **ExpenseRepository** class — each call to the **SaveChanges** method is a separate transaction, one of which may fail leading to the risk of orphaned detail records.

```csharp
C#
// Example code when using two tables for expenses data.
public void SaveExpense(Expense expense)
{
  // create an expense row.
  var context = new ExpenseDataContext(this.account);
  ExpenseRow expenseRow = expense.ToTableEntity();

  foreach (var expenseItem in expense.Details)
  {
    // create an expense item row.
    var expenseItemRow = expenseItem.ToTableEntity();
    expenseItemRow.PartitionKey = expenseRow.PartitionKey;
    expenseItemRow.RowKey =
      string.Format(CultureInfo.InvariantCulture, "{0}_{1}",
                    expense.Id, expenseItemRow.Id);

    context.AddObject(ExpenseDataContext.ExpenseItemTable,
                      expenseItemRow);
    ...
  }

  // save the expense item rows.
  cont ext.SaveChanges(SaveChangesOptions.Batch);

  // save the expense row.
  context.AddObject(ExpenseDataContext.ExpenseTable,
      expenseRow);
  context.SaveChanges();
  ...
}
```

To resolve this situation the developers would need to write code that implements a compensating transaction mechanism so that a failure when saving a header or detail row does not affect the integrity of the data. This is possible, but adds to the complexity of the solution. For example, to resolve the potential issue of orphaned detail records after a failure, the developers could implement an "orphan collector" process that will regularly scan the details table looking for, and deleting, orphaned records.

However, because the developers at Adatum chose to implement a multi-schema table for expense data, they can use a single transaction for saving both header and detail records. This approach enables them to use Entity Group Transactions to save an expense header entity and its related detail entities to a single partition in a single, atomic transaction. The following code sample from the Expense-Repository class shows how the application saves an expense to table storage.

Notice how the two overloaded versions of the **ToTableEntity** extension method return either an **IExpenseRow** or an **IExpenseItemRow** instance. The data access layer code is in the **DataAccessLayer** folder of the **aExpense.Shared** project in the example solution.

C#

```csharp
// Actual code used to save expenses data from a single table.
public void SaveExpense(Expense expense)
{
  var context = new ExpenseDataContext(this.account);
  IExpenseRow expenseRow = expense.ToTableEntity();
  expenseRow.PartitionKey = ExpenseRepository
      .EncodePartitionAndRowKey(expenseRow.UserName);
  expenseRow.RowKey = expense.Id.ToString();

  context.AddObject(ExpenseDataContext.ExpenseTable,
      expenseRow);

  foreach (var expenseItem in expense.Details)
  {
    // Create an expense item row.
    var expenseItemRow = expenseItem.ToTableEntity();
    expenseItemRow.PartitionKey = expenseRow.PartitionKey;
    expenseItemRow.RowKey = string.Format(
        CultureInfo.InvariantCulture, "{0}_{1}", expense.Id,
        expenseItemRow.ItemId);
    context.AddObject(ExpenseDataContext.ExpenseTable,
        expenseItemRow);

    // save receipt image if any
    if (expenseItem.Receipt != null
    && expenseItem.Receipt.Length > 0)
    {
      this.receiptStorage.AddReceipt(
          expenseItemRow.ItemId.ToString(),
          expenseItem.Receipt, string.Empty);
    }
  }

  // Save expense and the expense items row in the same
  // batch transaction using a retry policy.
  this.storageRetryPolicy.ExecuteAction(
    () => context.SaveChanges(SaveChangesOptions.Batch);
  ...
}
```

You can also see in the second example how Adatum chose to use the Enterprise Library Transient Fault Handling Application Block to retry the **SaveChanges** operation if it fails due to a temporary connectivity. The Windows Azure storage client API includes support for custom retry policies, but Adatum uses the Transient Fault Handling Application Block to take advantage of its customization capabilities and to implement a standard approach to all the retry logic in the application. See Chapter 4, "Moving to Windows Azure SQL Database," for information about using the Transient Fault Handling Application Block.

There are some additional restrictions on performing Entity Group Transactions: each entity can appear only once in the transaction, there must be no more than 100 entities in the transaction, and the total size of the request payload must not exceed 4 megabytes (MB). Adatum assumes that no one will submit more than 100 business expense items as part of a single submission, but will consider adding some additional validation to the application's code to prevent this.

Partition Keys and Row Keys

The second important decision about table storage is the selection of keys to use. Windows Azure table storage uses two keys: a partition key and a row key. Windows Azure uses the partition key to implement load balancing across storage nodes. The load balancer can identify "hot" partitions (partitions that contain data that is accessed more frequently than the data in other partitions) and run them on separate storage nodes in order to improve performance. This has deep implications for your data model design and your choice of partition keys:

• The partition key forms the first part of the tuple that uniquely identifies an entity in table storage. The row key is a unique identifier for an entity within a partition and forms the second part of the tuple that uniquely identifies an entity in table storage.

• You can only use Entity Group Transactions on entities in the same table and in the same partition. You may want to choose a partition key based on the transactional requirements of your application. Don't forget that a table can store multiple entity types.

Each entry in a table is simply a property bag. Each property bag can represent a different entity type; this means that a single partition can hold multiple entities of the same or different types.

- You can optimize queries based on your knowledge of partition keys. For example, if you know that all the entities you want to retrieve are located on the same partition, you can include the partition key in the **where** clause of the query. In a single query, accessing multiple entities from the same partition is much faster than accessing multiple entities on different partitions. If the entities you want to retrieve span multiple partitions, you can split your query into multiple queries and execute them in parallel across the different partitions.

> *If you want to create parallel queries, you should plan to use Parallel LINQ (PLINQ) instead of creating your own threads in the web role.*

Adatum determined that reverse chronological order is the most likely order in which the expense items will be accessed because users are typically interested in the most recent expenses. Therefore, it decided to use a row key that guarantees the expense items are stored in this order to avoid the need to sort them.

The following code sample from the **ExpenseKey** class shows how the static **Now** property generates an inverted tick count to use in its **InvertedTicks** property.

Choosing the right partition key is the most important decision you make that affects the performance of your storage solution. The partition key and row key together make up a tuple that uniquely identifies any entity in table storage.

```C#
public static ExpenseKey Now
{
  get
  {
    return new ExpenseKey(string.Format("{0:D19}",
        DateTime.MaxValue.Ticks - DateTime.UtcNow.Ticks));
  }
}
```

For the partition key, Adatum decided to use the **UserName** property because the vast majority of queries will filter based on a user name. For example, the website displays the expense submissions that belong to the logged on user.

This also enables the application to filter expense item rows by **ExpenseID** as if there was a foreign key relationship. The following code in the **SaveChanges** method in the **ExpenseRepository** class shows how the application creates this row key value for an expense item entity from the **Id** property of the expense header entity and the **Id** property of the expense item entity.

C#
```
expenseItemRow.RowKey = string.Format(
            CultureInfo.InvariantCulture,
              "{0}_{1}", expense.Id, expenseItemRow.Id);
```

The following code example shows how you could query for **Expense-Item** rows based on **ExpenseID** by including the partition key in the query.

C#
```
char charAfterSeparator =
      Convert.ToChar((Convert.ToInt32('_') + 1));
var nextId = expenseId.ToString() + charAfterSeparator;

var expenseItemQuery =
  (from expenseItem in context.ExpensesAndExpenseItems
   where
     expenseItem.RowKey.CompareTo(expenseId.ToString()) >= 0 &&
     expenseItem.RowKey.CompareTo(nextId) < 0 &&
     expenseItem.PartitionKey.CompareTo(expenseRow.PartitionKey)
       == 0
   select expenseItem).AsTableServiceQuery();
```

A more natural way of writing this query would be to use **StartsWith** instead of **CompareTo**. However, **StartsWith** is not supported by the Windows Azure table service. You also get performance benefits from this query because the **where** clause includes the partition key.

Windows Azure places some restrictions on the characters that you can use in partition and row keys. Generally speaking, the restricted characters are ones that are meaningful in a URL. For more information, see *"Understanding the Table Service Data Model."* In the aExpense application, it's possible that these illegal characters could appear in the **UserName** used as the partition key value for the Expense table.

> *If there is an illegal character in your partition key, Windows Azure will return a Bad Request (400) message.*

To avoid this problem, the aExpense application encodes the **User-Name** value using a base64 encoding scheme before using the **User-Name** value as a row key. Implementing base64 encoding and decoding is very easy.

```csharp
C#
public static string EncodePartitionAndRowKey(string key)
{
  if (key == null)
  {
    return null;
  }
  return Convert.ToBase64String(
      System.Text.Encoding.UTF8.GetBytes(key));
}

public static string DecodePartitionAndRowKey(string encodedKey)
{
  if (encodedKey == null)
  {
    return null;
  }
  return System.Text.Encoding.UTF8.GetString(
      Convert.FromBase64String(encodedKey));
}
```

The team at Adatum first tried to use the **UrlEncode** method because it would have produced a more human readable encoding, but this approach failed because it does not encode the percent sign (%) character.

> *According to the documentation, the percent sign character is not an illegal character in a key, but Adatum's testing showed that entities with a percent sign character in the key could not be deleted.*

Another approach would be to implement a custom escaping technique.

A custom method to transform the user name to a legal character sequence could leave the keys human-readable, which would be useful during debugging or troubleshooting.

Defining the Schemas

In the aExpense application, two types of entity are stored in the Expense table: expense header entities (defined by the **IExpenseRow** interface) and expense detail entities (defined by the **IExpenseItem-Row** interface). The following code sample shows these two interfaces and the **IRow** interface that defines the entity key.

C#
```csharp
public interface IExpenseRow : IRow
{
  // NOTE: DateTime bool and Guid types must be Nullable
  // in order to run in the storage emulator.
  string Id { get; set; }
  string UserName { get; set; }
  bool? Approved { get; set; }
  string ApproverName { get; set; }
  string CostCenter { get; set; }
  DateTime? Date { get; set; }
  string ReimbursementMethod { get; set; }
  string Title { get; set; }
}

public interface IExpenseItemRow : IRow
{
  Guid? ItemId { get; set; }
  string Description { get; set; }
  double? Amount { get; set; }
  string ReceiptUrl { get; set; }
  string ReceiptThumbnailUrl { get; set; }
}

public interface IRow
{
  string PartitionKey { get; set; }
  string RowKey { get; set; }
  DateTime Timestamp { get; set; }
  string Kind { get; set; }
}
```

> Both entity types share the same key structure defined in the **IRow** interface.

Adatum had to make a change to the data type that the application uses to store the business expense amount. In Windows Azure SQL Database, this field was stored as a decimal. This data type is not supported in Windows Azure table storage and the amount is now stored as a double.

Adatum uses the **ExpenseAndExpenseItemRow** and **Row** classes to implement the **IRow, IExpense-Row**, and **IExpenseItemRow** interfaces, and to extend the **TableServiceEntity** class from the **Storage-Client** namespace. The following code sample shows the **Row** and **ExpenseAndExpenseItemRow** classes. The **Row** class defines a **Kind** property that is used to distinguish between the two types of entity stored in the table (see the **TableRows** enumeration in the **DataAccessLayer** folder of the **aExpense.Shared** project).

```csharp
C#
public abstract class Row : TableServiceEntity, IRow
{
  protected Row()
  { }

  protected Row(string kind) : this(null, null, kind)
  { }

  protected Row(
      string partitionKey, string rowKey, string kind)
      : base(partitionKey, rowKey)
  {
    this.Kind = kind;
  }

  public string Kind { get; set; }
}

public class ExpenseAndExpenseItemRow
    : Row, IExpenseRow, IExpenseItemRow
{
  public ExpenseAndExpenseItemRow()
  { }

  public ExpenseAndExpenseItemRow(TableRows rowKind)
  {
    this.Kind = rowKind.ToString();
  }

  // Properties from ExpenseRow
  public string Id { get; set; }
  public string UserName { get; set; }
  public bool? Approved { get; set; }
  public string ApproverName { get; set; }
  public string CostCenter { get; set; }
  public DateTime? Date { get; set; }
  public string ReimbursementMethod { get; set; }
  public string Title { get; set; }
```

```
    // Properties from ExpenseItemRow
    public Guid? ItemId { get; set; }
    public string Description { get; set; }
    public double? Amount { get; set; }
    public string ReceiptUrl { get; set; }
    public string ReceiptThumbnailUrl { get; set; }
}
```

The following code example shows how the **ExpenseDataContext** class maps the **ExpenseAndExpenseItemRow** class to a Windows Azure storage table named **multientityschemaexpenses**.

```
C#
public class ExpenseDataContext : TableServiceContext
{
  public const string ExpenseTable = "multientityschemaexpenses";

  ...

  public IQueryable<ExpenseAndExpenseItemRow>
        ExpensesAndExpenseItems
  {
    get
    {
      return this.CreateQuery<ExpenseAndExpenseItemRow>(
            ExpenseTable);
    }
  }

  ...
}
```

Retrieving Records from a Multi-Entity Schema Table

Storing multiple entity types in the same table does add to the complexity of the application. The aExpense application uses LINQ to specify what data to retrieve from table storage. The following code example shows how the application retrieves expense submissions for approval by approver name.

*Use the **AsTableServiceQuery** method to return data from Windows Azure table storage.*

```
C#
var query = (from expense in context.ExpensesAndExpenseItems
     where expense.ApproverName.CompareTo(approverName) == 0
   select expense).AsTableServiceQuery();
return this.storageRetryPolicy.ExecuteAction(
          () => query.Execute()).SingleOrDefault();
```

The **AsTableServiceQuery** method converts the standard **IQueryable** result to a **CloudTableQuery** result. Using a **CloudTableQuery** object offers the following benefits to the application:

- Data can be retrieved from the table in multiple segments instead of getting it all in one go. This is useful when dealing with a large set of data.
- You can specify a retry policy for cases when the query fails. However, as you saw earlier, Adatum chose to use the Transient Fault Handling Block instead.

The query methods in the **ExpenseRepository** class use the **Expense-AndExpenseItemRow** entity class when they retrieve either header or detail entities from the expense table. The following code example from the **GetExpensesByUser** method in the **ExpenseRespository** class shows how to retrieve a header row (defined by the **IExpense-Row** interface).

The use of the **Take** method is not intended as a paging mechanism. It is included in order to improve the performance of the code. Using the partition key in the query improves the performance because the partition key is indexed. This example does not need to use the **Kind** property because only header entities have a **UserName** property.

```csharp
C#
var context = new ExpenseDataContext(this.account)
                { MergeOption = MergeOption.NoTracking };

var query = (from expense in context.ExpensesAndExpenseItems
            where
                expense.UserName.CompareTo(userName) == 0 &&
                expense.PartitionKey.CompareTo(
                    EncodePartitionAndRowKey(userName)) == 0
            select expense).Take(10).AsTableServiceQuery();

try
{
  return this.storageRetryPolicy.ExecuteAction(
      () => query.Execute()).Select(e => e.ToModel()).ToList();
}
...
```

The following code sample from the **GetExpensesById** method in
the **ExpenseRepository** class uses the **Kind** property to select only
detail entities.

```C#
var expenseAndItemRows = query.Execute().ToList();
...
expenseAndItemRows.
    Where(e => e.Kind == TableRows.ExpenseItem.ToString()).
    Select(e => (e as IExpenseItemRow).ToModel()).
    ToList().ForEach(e => expense.Details.Add(e));
```

Try to handle all data access
issues within your data
access layer.

Materializing Entities

In the aExpense application, all the methods in the **ExpenseRepository**
class that return data from queries call the **ToList** method before re-
turning the results to the caller.

```C#
public IEnumerable<Expense> GetExpensesForApproval(string
approverName)
{
  ExpenseDataContext context = new
        ExpenseDataContext(this.account);

  var query = (from expense in context.ExpensesAndExpenseItems
          where
            expense.ApproverName.CompareTo(approverName) == 0
          select expense).AsTableServiceQuery();

  try
  {
    return this.storageRetryPolicy.ExecuteAction(() =>
          query.Execute()).Select(e => e.ToModel()).ToList();
  }
  catch (InvalidOperationException)
  {
    Log.Write(EventKind.Error,
      "By calling ToList(), this exception can be handled
       inside the repository.");
    throw;
  }
}
```

The reason for this is that calling the **Execute** method does not materialize the entities. Materialization does not happen until someone calls **MoveNext** on the **IEnumerable** collection. Without **ToList**, the first call to **MoveNext** happens outside the repository. The advantage of having the first call to the **MoveNext** method inside the **Expense-Repository** class is that you can handle any data access exceptions inside the repository.

Query Performance

As mentioned earlier, the choice of partition key can have a big impact on the performance of the application. This is because Windows Azure tracks activity at the partition level, and can automatically migrate a busy partition to a separate storage node in order to improve data access performance for the application.

Adatum uses partition keys in queries to improve the performance. For example, the following query to retrieve stored business expense submissions for a user by using this query would work, even though it does not specify a partition key.

```csharp
C#
var query = (from expense in context.ExpensesAndExpenseItems
                where
                    expense.UserName.CompareTo(userName) == 0
                select expense).AsTableServiceQuery();
```

It's important to understand the impact that partitions can have on query performance.

However, this query must scan all the partitions of the table to search for matching records. This is inefficient if there are a large number of records to search, and its performance may be further affected if it has to scan data across multiple storage nodes sequentially.

Adatum's test team did performance testing on the application using queries that do not include the partition key, and then evaluated the improvement when the partition key is included in the **where** clause. The testers found that there was a significant performance improvement in the aExpense application using a query that includes the partition key, as shown here.

C#
```
var query = (from expense in context.ExpensesAndExpenseItems
            where
              expense.UserName.CompareTo(userName) == 0
              && expense.PartitionKey.CompareTo(
                EncodePartitionAndRowKey(userName)) == 0
            select expense).Take(10).AsTableServiceQuery();
```

*If a table query does not include the partition key in its **where** clause, you should re-evaluate your choice of row key and partition key for the table to avoid the potential performance problems associated with scanning multiple partitions.*

Working with Development Storage

There are some differences between development table storage and Windows Azure table storage documented at *"Differences Between the Storage Emulator and Windows Azure Storage Services."* The team at Adatum encountered the error "One of the request inputs is not valid" that occurs when testing the application with empty tables in development storage.

The solution that Adatum adopted was to insert, and then delete, a dummy row into the Windows Azure tables if the application is using the local storage emulator. During the initialization of the web role, the application calls the **CreateTableIfNotExist<T>** extension method in the **TableStorageExtensionMethods** class to check whether it is running against local development storage. If this is the case it adds and then deletes a dummy record in the application's Windows Azure tables.

The following code from the **TableStorageExtensionMethods** class (defined in the **Source\Shared\aExpense** folder) demonstrates how the aExpense application determines whether it is using development storage and how it adds and deletes a dummy record to the table.

Don't assume that local development storage will work in exactly the same way as Windows Azure storage. You should consider adding dummy records to all tables in local development storage.

```csharp
C#
public static bool CreateTableIfNotExist<T>(
    this CloudTableClient tableStorage, string entityName)
    where T : TableServiceEntity, new()
{
  bool result = tableStorage.CreateTableIfNotExist(entityName);

  // Execute conditionally for development storage only
  if (tableStorage.BaseUri.IsLoopback)
  {
    InitializeTableSchemaFromEntity(tableStorage,
        entityName, new T());
  }
  return result;
}

private static void InitializeTableSchemaFromEntity(
    CloudTableClient tableStorage, string entityName,
    TableServiceEntity entity)
{
  TableServiceContext context =
        tableStorage.GetDataServiceContext();
  DateTime now = DateTime.UtcNow;
  entity.PartitionKey = Guid.NewGuid().ToString();
  entity.RowKey = Guid.NewGuid().ToString();
  Array.ForEach(
    entity.GetType().GetProperties(BindingFlags.Public |
    BindingFlags.Instance),
    p =>
    {
      if ((p.Name != "PartitionKey") &&
          (p.Name != "RowKey") && (p.Name != "Timestamp"))
      {
        if (p.PropertyType == typeof(string))
        {
          p.SetValue(entity, Guid.NewGuid().ToString(), null);
        }
        else if (p.PropertyType == typeof(DateTime))
        {
          p.SetValue(entity, now, null);
        }
      }
    });

  context.AddObject(entityName, entity);
  context.SaveChangesWithRetries();
  context.DeleteObject(entity);
  context.SaveChangesWithRetries();
}
```

Storing Profile Data

Until now Adatum has used the built-in ASP.NET profile mechanism to store each user's preferred reimbursement method. In Windows Azure, the ASP.NET profile provider communicates with either SQL Server or Windows Azure SQL Database (depending on the previous migration stage) where the ASPNETDB database resides. However, during this final migration step Adatum will move away from using a relational database in favor of storing all of the application data in Windows Azure table and blob storage. Therefore it makes no sense to continue to use a relational database just for the profile data.

Instead, Adatum chose to use a sample provider that utilizes Windows Azure table storage to store profile information. You can download this provider from *"Windows Azure ASP.NET Providers Sample."* The only change required for the application to use a different profile provider is in the Web.config file.

Using a profile provider to access profile data minimizes the code changes required in the application.

```XML
<profile defaultProvider="TableStorageProfileProvider">
  <providers>
    <clear />
    <add name="TableStorageProfileProvider"
         type="AExpense.Providers.TableStorageProfileProvider …"
         applicationName="aExpenseProfiles" />
  </providers>

  <properties>
    <add name="PreferredReimbursementMethod" />
  </properties>
</profile>
```

Using the **TableStorageProfileProvider** class does raise some issues for the application:

- The table storage profile provider is unsupported sample code.

- You must migrate your existing profile data from SQL Server to Windows Azure table storage.

    *In the example provided for this guide, the table is populated by the **Initialize** method of the **ProfileInitializer** class defined in the **WebRole** class, which is executed when the application starts. In a real application, users would have the ability to set their own preferences.*
    Adatum should also migrate users existing preferences from the SQL Database tables to Windows Azure table storage.

- You need to consider whether, in the long run, Windows Azure table storage is suitable for storing profile data.

Even with these considerations to taken into account, using the table storage profile provider enabled Adatum to get rid of the need for a relational database; which helps to minimize the running costs of the application.

> *Chapter 6, "Evaluating Cloud Hosting Costs," describes the relative costs of using Windows Azure storage and Windows Azure SQL Database.*

Generating and Exporting the Expense Data

The developers at Adatum added functionality to the aExpense application to export a summary of the approved expenses data to use with Adatum's existing on-premises reimbursement system.

Generating the Expense Report Table

The task that performs this operation uses the worker role plumbing code described in Chapter 5, "Executing Background Tasks." The discussion here will focus on the task implementation and table design issues; it does not focus on the plumbing code.

This task is the first of two tasks that generate the approved expense data for export. It is responsible for generating the "flattened" table of approved expense data in Windows Azure table storage. The following code sample shows how the expense report export process begins in the **ExpenseRepository** class (in the **DataAccessLayer** folder of the **aExpense.Shared** project) where the **UpdateApproved** method adds a message to a queue and updates the **Approved** property of the expense header record.

For this task, we were able to use our worker role plumbing code without modification.

```C#
public void UpdateApproved(Expense expense)
{
  var context = new ExpenseDataContext(this.account);

  ExpenseRow expenseRow =
      GetExpenseRowById(context, expense.Id);
  expenseRow.Approved = expense.Approved;

  var queue = new AzureQueueContext(this.account);
  this.storageRetryPolicy.ExecuteAction(
    () => queue.AddMessage(new ApprovedExpenseMessage {
                ExpenseId = expense.Id.ToString(),
                ApproveDate = DateTime.UtcNow }));

  context.UpdateObject(expenseRow);
  this.storageRetryPolicy.ExecuteAction(
    () => context.SaveChanges());
}
```

This code uses a new message type named **ApprovedExpenseMessage** that derives from the plumbing code class named **BaseQueueMessage**. The following code sample shows the two properties of the **ApprovedExpenseMessage** class.

```C#
[DataContract]
public class ApprovedExpenseMessage : BaseQueueMessage
{
  [DataMember]
  public string ExpenseId { get; set; }

  [DataMember]
  public DateTime ApproveDate { get; set; }
}
```

> We "flatten" the data and calculate the expense submission total before saving the data into an intermediate table. This table contains the data structured in exactly the format we need to export.

The following code shows how the **ProcessMessage** method in the **ExpenseExportJob** class (located in the **Jobs** folder of the **aExpense. Workers** project) retrieves the message from the queue and creates a new **ExpenseExport** entity to save to table storage.

```csharp
C#
public override bool ProcessMessage(
    ApprovedExpenseMessage message)
{
  try
  {
    Expense expense = this.expenses.GetExpenseById(
            new ExpenseKey(message.ExpenseId));

    if (expense == null)
    {
      return false;
    }

    // If the expense was not updated but a message was
    // persisted, we need to delete it.
    if (!expense.Approved)
    {
      return true;
    }

    double totalToPay = expense.Details.Sum(x => x.Amount);
    var export = new ExpenseExport
      {
        ApproveDate = message.ApproveDate,
        ApproverName = expense.ApproverName,
        CostCenter = expense.CostCenter,
        ExpenseId = expense.Id,
        ReimbursementMethod = expense.ReimbursementMethod,
        TotalAmount = totalToPay,
        UserName = expense.User.UserName
      };
    this.expenseExports.Save(export);
  }
  catch (InvalidOperationException ex)
```

```
  {
    var innerEx =
          ex.InnerException as DataServiceClientException;
    if (innerEx != null &&
        innerEx.StatusCode == (int)HttpStatusCode.Conflict)
    {
      // The data already exists, so we can return true
      // because we have processed this before.
      return true;
    }
    Log.Write(EventKind.Error, ex.TraceInformation());
    return false;
  }

  return true;
}
```

If this method fails for any reason other than a conflict on the insert, the plumbing code classes ensure that message is left on the queue. When the **ProcessMessage** method tries to process the message from the queue a second time, the insert to the expense report table fails with a duplicate key error and the inner exception reports this as a conflict in its **StatusCode** property. If this happens, the method can safely return a true result.

If the **Approved** property of the **Expense** object is false, this indicates a failure during the **UpdateApproved** method after it added a message to the queue, but before it updated the table. In this circumstance, the **ProcessMessage** method removes the message from the queue without processing it.

The partition key of the Expense Export table is the expense approval date, and the row key is the expense ID. This optimizes access to this data for queries that use the approval date in the **where** clause, which is what the export process requires.

Exporting the Expenses Data

This task is the second of two tasks that generate the approved expense data for export. It is responsible for creating a Windows Azure blob that contains a CSV file of approved expense submissions data.

We need to ensure that this process is robust. We don't want to lose any expense submissions, or pay anyone twice.

Choose partition keys and rows keys to optimize your queries against the data. Ideally, you should be able to include the partition key in the **where** block of the query.

The worker role plumbing code classes now support scheduled tasks in addition to tasks that are triggered by a message on a queue.

The task that generates the blob containing the expense report data is slightly different from the two other tasks in the aExpense application. The other tasks poll a queue to see if there is any work for them to do. The export task is triggered by a schedule, which sets the task to run at fixed times. The team at Adatum had to modify their worker role plumbing code classes to support scheduled tasks.

You can use the abstract class **JobProcessor**, which implements the **IJobProcessor** interface, to define new scheduled tasks. The following code example shows the **JobProcessor** class.

We could extend the application to enable an on-premises application to generate an ad-hoc expense data report by allowing an on-premises application to place a message onto a Windows Azure queue. We could then have a task that generated the report data when it received a message on the queue.

```csharp
C#
public abstract class JobProcessor : IJobProcessor
{
  private bool keepRunning;

  protected JobProcessor(int sleepInterval)
  {
    if (sleepInterval <= 0)
    {
      throw new ArgumentOutOfRangeException("sleepInterval");
    }

    this.SleepInterval = sleepInterval;
  }

  protected int SleepInterval { get; set; }

  public void Run()
  {
    this.keepRunning = true;
    while (this.keepRunning)
    {
      Thread.Sleep(this.SleepInterval);
      this.RunCore();
    }
  }

  public void Stop()
  {
    this.keepRunning = false;
  }

  protected abstract void RunCore();
}
```

This implementation does not make it easy to specify the exact time that scheduled tasks will run. The time between tasks will be the value of the sleep interval, plus the time taken to run the task. If you need the task to run at a fixed time, you should measure how long the task takes to run and subtract that value from the sleep interval.

> *The **BaseJobProcessor** class that defines tasks that read messages from queues extends the **JobProcessor** class.*

In the aExpense application, the **ExpenseExportBuilderJob** class extends the **JobProcessor** class to define a scheduled task. The **ExpenseExportBuilderJob** class, shown in the following code example, defines the task that generates the expense report data and stores it as a blob. In this class, the **expenseExports** variable refers to the table of approved expense submissions, and the **exportStorage** variable refers to the report data in blob storage that will be downloaded. The call to the base class constructor specifies the interval at which the job runs.

> *The following code sets the scheduled interval to a low number for testing and demonstration purposes. You should change this interval for a "real" schedule.*

```C#
public class ExpenseExportBuilderJob : JobProcessor
{
  private readonly ExpenseExportRepository expenseExports;
  private readonly ExpenseExportStorage exportStorage;

  public ExpenseExportBuilderJob() : base(100000)
  {
    this.expenseExports = new ExpenseExportRepository();
    this.exportStorage = new ExpenseExportStorage();
  }
```

In the **RunCore** method, the code first retrieves all the approved expense submissions from the export table based on the job date. Next, the code appends a CSV record to the export data in blob storage for each approved expense submission. Finally, the code deletes from the table all the records it copied to blob storage.

```csharp
C#
  protected override void RunCore()
  {
    DateTime jobDate = DateTime.UtcNow;
    string name = jobDate.ToExpenseExportKey();

    IEnumerable<ExpenseExport> exports =
          this.expenseExports.Retreive(jobDate);
    if (exports == null || exports.Count() == 0)
    {
      return;
    }

    string text = this.exportStorage.GetExport(name);
    var exportText = new StringBuilder(text);
    foreach (ExpenseExport expenseExport in exports)
    {
      exportText.AppendLine(expenseExport.ToCsvLine());
    }

    this.exportStorage.AddExport(name,
        exportText.ToString(), "text/plain");

    // Delete the exports.
    foreach (ExpenseExport exportToDelete in exports)
    {
      try
      {
        this.expenseExports.Delete(exportToDelete);
      }
      catch (InvalidOperationException ex)
      {
        Log.Write(EventKind.Error, ex.TraceInformation());
      }
    }
  }
}
```

If the process fails before it deletes all the approved expense submissions from the export table, any undeleted approved expense submissions will be exported a second time when the task next runs. However, the exported CSV data includes the expense ID and the approval date of the expense submission, so the on-premises payment processing system will be able to identify duplicate items.

The following code shows the methods that the **RunCore** method invokes to retrieve approved expense submissions and delete them after it copies them to the export blob. These methods are defined in the **ExpenseExportRepoisitory** class located in the **DataAccessLayer** folder of the **aExpense. Shared** project. Because they use the job date to identify the partitions to search, these queries are fast and efficient.

```csharp
C#
public IEnumerable<ExpenseExport> Retreive(DateTime jobDate)
{
  var context = new ExpenseDataContext(this.account);
  string compareDate = jobDate.ToExpenseExportKey();
  var query = (from export in context.ExpenseExport
          where export.PartitionKey.CompareTo(compareDate) <= 0
          select export).AsTableServiceQuery();

  var val = query.Execute();
  return val.Select(e => e.ToModel()).ToList();
}

public void Delete(ExpenseExport expenseExport)
{
  var context = new ExpenseDataContext(this.account);
  var query = (from export in context.ExpenseExport
      where export.PartitionKey.CompareTo(
        expenseExport.ApproveDate.ToExpenseExportKey()) == 0 &&
        export.RowKey.CompareTo(
        expenseExport.ExpenseId.ToString()) == 0
      select export).AsTableServiceQuery();
  ExpenseExportRow row = query.Execute().SingleOrDefault();
  if (row == null)
  {
    return;
  }

  context.DeleteObject(row);
  context.SaveChanges();
}
```

PERFORMANCE TESTING, TUNING, TO-DO ITEMS

Adatum made changes to the aExpense application following performance testing.

As part of the work for this phase, the team at Adatum evaluated the results from performance testing the application and, as a result, made a number of changes to the aExpense application. They also documented some of the key "missing pieces" in the application that Adatum should address in the next phase of the project.

Initializing the Storage Tables, Blobs, and Queues

During testing of the application, the team at Adatum discovered that the code that creates the expenses storage repository and the job that processes receipt images were affecting performance. They isolated this to the fact that the code calls the **CreateIfNotExist** method every time the repository is instantiated, which requires a round-trip to the storage server to check whether the receipt container exists. This also incurs an unnecessary storage transaction cost. To resolve this, the developers realized that they should create the receipt container only once when the application starts.

Originally, the constructor for the **ExpenseReceiptStorage** class was responsible for checking that the expense receipt container existed, and creating it if necessary. This constructor is invoked whenever the application instantiates an **ExpenseRepository** object or a **Receipt-ThumbnailJob** object. The **CreateIfNotExist** method that checks whether a container exists requires a round-trip to the storage server and incurs a storage transaction cost.

To avoid these unnecessary round-trips, Adatum moved this logic to the **ApplicationStorageInitializer** class defined in the **WebRole** class. This class prepares all of the tables, blobs, and queues required by the application when the role first starts.

```csharp
C#
public static class ApplicationStorageInitializer
{
  public static void Initialize()
  {
    CloudStorageAccount account =
      CloudConfiguration.GetStorageAccount(
                          "DataConnectionString");
```

```
    // Tables - create if they do not already exist.
    var cloudTableClient =
      new CloudTableClient(account.TableEndpoint.ToString(),
                           account.Credentials);
      cloudTableClient.CreateTableIfNotExist<
                       ExpenseAndExpenseItemRow>(
                           ExpenseDataContext.ExpenseTable);
      cloudTableClient.CreateTableIfNotExist<ExpenseExportRow>(
                       ExpenseDataContext.ExpenseExportTable);

    // Blobs - create if they do not already exist.
    var client = account.CreateCloudBlobClient();
    client.RetryPolicy = RetryPolicies.Retry(3,
                           TimeSpan.FromSeconds(5));
    var container = client.GetContainerReference(
                    ExpenseReceiptStorage.ReceiptContainerName);
    container.CreateIfNotExist();
    container = client.GetContainerReference(
            ExpenseExportStorage.ExpenseExportContainerName);
    container.CreateIfNotExist();

    // Queues - remove any existing stored messages
    var queueContext = new AzureQueueContext(account);
    queueContext.Purge<NewReceiptMessage>();
    queueContext.Purge<ApprovedExpenseMessage>();
  }
}
```

The **Application_Start** method in the Global.asax.cs file and the **OnStart** method of the worker role invoke the **Initialize** method in this class.

> *You may find that you can improve performance when making small requests to Windows Azure storage queues, tables, and blobs by changing the service point settings. See "Nagle's Algorithm is Not Friendly towards Small Requests" for more information.*

Implementing Paging with Windows Azure Table Storage

During performance testing, the response times for Default.aspx degraded as the test script added more and more expense submissions for a user. This happened because the current version of the Default.aspx page does not include any paging mechanism, so it always displays all the expense submissions for a user. As a temporary measure, Adatum modified the LINQ query that retrieves expense submissions by user to include a **Take(10)** clause, so that the application only requests the first 10 expense submissions. In a future phase of the project, Adatum will add paging functionality to the Default.aspx page.

Adatum has not implemented any paging functionality in the current phase of the project, but this section gives an outline of the approach it intends to take. The **ResultSegment** class in the Windows Azure **StorageClient** library provides an opaque **ContinuationToken** property that you can use to access the next set of results from a query if that query did not return the full set of results; for example, if the query used the **Take** operator to return a small number of results to display on a page. This **ContinuationToken** property will form the basis of any paging implementation.

The **ResultSegment** class only returns a **ContinuationToken** object to access the next page of results, and not the previous page, so if your application requires the ability to page backward, you must store **ContinuationToken** objects that point to previous pages. A stack is a suitable data structure to use. Figure 7 shows the state of a stack after a user has browsed to the first page and then paged forward as far as the third page.

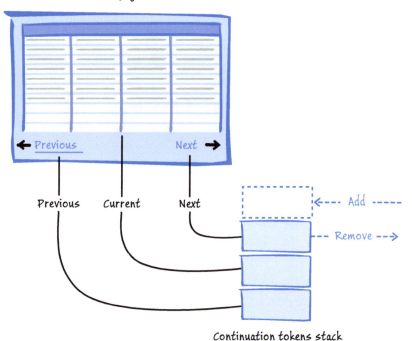

FIGURE 7
Displaying page 3 of the data from a table

If a user clicks the **Next** hyperlink to browse to page 4, the page peeks at the stack to get the continuation token for page 4. After the page executes the query with the continuation token from the stack, it pushes a new continuation token for page 5 onto the stack.

If a user clicks the **Previous** hyperlink to browse to page 2, the page will pop two entries from the stack, and then peek at the stack to get the continuation token for page 2. After the page executes the query with the continuation token from the stack, it will push a new continuation token for page 3 onto the stack.

The following code examples show how Adatum could implement this behavior in an asynchronous ASP.NET page.

> *Using an asynchronous page frees up the pages thread from the thread pool while a potentially long-running I/O operation takes place. This improves throughput on the web server and increases the scalability of the application.*

The following two code examples show how to create an asynchronous ASP.NET page. First, add an **Async="true"** attribute to the page directive in the .aspx file.

```HTML
<%@ Page Language="C#" AutoEventWireup="true"
        CodeBehind="Default.aspx.cs"
        Inherits="ContinuationSpike._Default"
        Async="true"%>
```

Second, register begin and end methods for the asynchronous operation in the load event for the page.

```C#
protected void Page_Load(object sender, EventArgs e)
{
  AddOnPreRenderCompleteAsync(
      new BeginEventHandler(BeginAsyncOperation),
      new EndEventHandler(EndAsyncOperation)
  );
}
```

We need to store the stack containing the continuation tokens as a part of the session state.

The following code example shows the definition of the **Continuation-Stack** class that the application uses to store continuation tokens in the session state.

```csharp
C#
public class ContinuationStack
{
  private readonly Stack stack;

  public ContinuationStack()
  {
    this.stack = new Stack();
  }

  public bool CanMoveBack()
  {
    if (this.stack.Count >= 2) return true;
    return false;
  }

  public bool CanMoveForward()
  {
    return this.GetForwardToken() != null;
  }

  public ResultContinuation GetBackToken()
  {
    if (this.stack.Count == 0) return null;
    // We need to pop twice and then return the next token.
    this.stack.Pop();
    this.stack.Pop();
    if (this.stack.Count == 0) return null;
    return this.stack.Peek() as ResultContinuation;
  }

  public ResultContinuation GetForwardToken()
  {
    if (this.stack.Count == 0) return null;
    return this.stack.Peek() as ResultContinuation;
  }

  public void AddToken(ResultContinuation result)
  {
    this.stack.Push(result);
  }
}
```

The following code example shows the **BeginAsyncOperation** method that starts the query execution for the next page of data. The **ct** value in the query string specifies the direction to move.

```C#
private IAsyncResult BeginAsyncOperation(object sender, EventArgs
e, AsyncCallback cb, object extradata)
{
  var query =
    new MessageContext(CloudConfiguration.GetStorageAccount())
      .Messages.Take(3).AsTableServiceQuery();
  if (Request["ct"] == "forward")
  {
    var segment = this.ContinuationStack.GetForwardToken();
    return query.BeginExecuteSegmented(segment, cb, query);
  }

  if (Request["ct"] == "back")
  {
    var segment = this.ContinuationStack.GetBackToken();
    return query.BeginExecuteSegmented(segment, cb, query);
  }
  return query.BeginExecuteSegmented(cb, query);
}
```

The **EndAsyncOperation** method puts the query results into the messages list and pushes the new continuation token onto the stack.

```C#
private List<MessageEntity> messages;

private void EndAsyncOperation(IAsyncResult result)
{
  var cloudTableQuery =
        result.AsyncState as CloudTableQuery<MessageEntity>;
  ResultSegment<MessageEntity> resultSegment =
        cloudTableQuery.EndExecuteSegmented(result);
  this.ContinuationStack.AddToken(
        resultSegment.ContinuationToken);
  this.messages = resultSegment.Results.ToList();
}
```

Preventing Users from Uploading Large Images

To prevent users from uploading large images of receipt scans to aExpense, Adatum configured the application to allow a maximum upload size of 1,024 kilobytes (KB) to the AddExpense.aspx page. The following code example shows the setting in the Web.config file.

```XML
<location path="AddExpense.aspx">
  <system.web>
    <authorization>
      <allow roles="Employee" />
      <deny users="*"/>
    </authorization>

    <!--
      Maximum request allowed to send a big image as a receipt.
      -->
    <httpRuntime maxRequestLength="1024"/>
  </system.web>
</location>
```

Validating User Input

The cloud-based version of aExpense does not perform comprehensive checks on user input for invalid or dangerous items. The AddExpense.aspx file includes some basic validation that checks the length of user input, but Adatum should add additional validation checks to the **OnAddNewExpenseItemClick** method in the AddExpense.aspx.cs file.

System.Net Configuration Changes

The following code example shows two configuration changes that Adatum made to the aExpense application to improve its performance.

```XML
<system.net>
  <settings>
    <servicePointManager expect100Continue="false" />
  </settings>
  <connectionManagement>
    <add address = "*" maxconnection = "24" />
  </connectionManagement>
</system.net>
```

The first change switches off the "Expect 100-continue" feature. If this feature is enabled, when the application sends a PUT or POST request, it can delay sending the payload by sending an "Expect 100-continue" header. When the server receives this message, it uses the available information in the header to check whether it could make the call, and if it can, it sends back a status code 100 to the client. The client then sends the remainder of the payload. This means that the client can check for many common errors without sending the payload.

If you have tested the client well enough to ensure that it is not sending any bad requests, you can turn off the "Expect 100-continue" feature and reduce the number of round trips to the server. This is especially useful when the client sends many messages with small payloads; for example, when the client is using the table or queue service.

The second configuration change increases the maximum number of connections that the web server will maintain from its default value of two. If this value is set too low, the problem manifests itself through "Underlying connection was closed" messages.

> *The exact number to use for this setting depends on your application. The page "Contention, poor performance, and deadlocks when you make Web service requests from ASP.NET applications" has useful information about how to set this for server side applications. You can also set it for a particular URI by specifying the URI in place of "*".*

WCF Data Service Optimizations

Because of a known performance issue with WCF Data Services, Adatum defined a **ResolveType** delegate on the **ExpenseDataContext** class in the aExpense application. Without this delegate, query performance degrades as the number of entities that the query returns increases. The following code example shows the delegate definition.

We made a number of changes to our WCF Data Services code to improve performance.

```csharp
private static Type ResolveEntityType(string name)
{
  var tableName = name.Split(new[] { '.' }).Last();
  switch (tableName)
  {
    case ExpenseTable:
      return typeof(ExpenseRow);
    case ExpenseItemTable:
      return typeof(ExpenseItemRow);
    case ExpenseExportTable:
      return typeof(ExpenseExportRow);
  }

  throw new ArgumentException(
      string.Format(
          CultureInfo.InvariantCulture,
          "Could not resolve the table name '{0}'
          to a known entity type.", name));
}
```

*Instead of using the **ResolveType** delegate, you can avoid the performance problem by ensuring that your entity class names exactly match the table names.*

Adatum added a further optimization to the WCF Data Services client code by setting the **Merge-Option** to **NoTracking** for the queries in the **ExpenseRepository** class. If you are not making any changes to the entities that WCF Data Services retrieve, there is no need for the **DataContext** object to initialize change tracking for entities.

MORE INFORMATION

All links in this book are accessible from the book's online bibliography available at: *http://msdn.microsoft.com/en-us/library/ff803373.aspx*.

"Blobs, Queues, and Tables" discusses the use of Windows Azure blobs, tables, and queues.

"Data Management" explores the options for storing data in Windows Azure SQL Database and blob storage.

The *Windows Azure Managed Library* includes detailed reference information for the **Microsoft. WindowsAzure.StorageClient** namespace.

"Windows Azure Storage Services REST API Reference" explains how you can interact with Windows Azure storage using scripts and code.

Glossary

affinity group. A named grouping that is in a single data center. It can include all the components associated with an application, such as storage, Windows Azure SQL Database instances, and roles.

autoscaling. Automatically scaling an application based on a schedule or on metrics collected from the environment.

claim. A statement about a subject; for example, a name, identity, key, group, permission, or capability made by one subject about itself or another subject. Claims are given one or more values and then packaged in security tokens that are distributed by the issuer.

cloud. A set of interconnected servers located in one or more data centers.

code near. When an application and its associated database(s) are both in the cloud.

code far. When an application is on-premises and its associated database(s) are in the cloud.

compute emulator. The Windows Azure compute emulator enables you to run, test, debug, and fine-tune your application before you deploy it as a hosted service to Windows Azure. See also: storage emulator.

Content Delivery Network (CDN). A system composed of multiple servers that contain copies of data. These servers are located in different geographical areas so that users can access the copy that is closest to them.

Enterprise Library. A collection of reusable software components (application blocks) designed to assist software developers with common enterprise development cross-cutting concerns (such as logging, validation, data access, exception handling, and many others).

horizontal scalability. The ability to add more servers that are copies of existing servers.

hosted service. Spaces where applications are deployed.

idempotent operation. An operation that can be performed multiple times without changing the result. An example is setting a variable.

Infrastructure as a Service (IaaS). A collection of infrastructure services such as storage, computing resources, and network that you can rent from an external partner.

lease. An exclusive write lock on a blob that lasts until the lease expires.

optimistic concurrency. A concurrency control method that assumes that multiple changes to data can complete without affecting each other; therefore, there is no need to lock the data resources. Optimistic concurrency assumes that concurrency violations occur infrequently and simply disallows any updates or deletions that cause a concurrency violation.

Platform as a Service (Paas). A collection of platform services that you can rent from an external partner that enable you to deploy and run your application without the need to manage any infrastructure.

poison message. A message that contains malformed data that causes the queue processor to throw an exception. The result is that the message isn't processed, stays in the queue, and the next attempt to process it once again fails.

Representational State Transfer (REST). An architectural style for retrieving information from websites. A resource is the source of specific information. Each resource is identified by a global identifier, such as a Uniform Resource Identifier (URI) in HTTP. The representation is the actual document that conveys the information.

service configuration file. Sets values for the service that can be configured while the hosted service is running. The values you can specify in the service configuration file include the number of instances that you want to deploy for each role, the values for the configuration parameters that you established in the service definition file, and the thumbprints for any SSL certificates associated with the service.

service definition file. Defines the roles that comprise a service, optional local storage resources, configuration settings, and certificates for SSL endpoints.

service package. Packages the role binaries and service definition file for publication to the Windows Azure Cloud Services.

snapshot. A read-only copy of a blob.

Storage Emulator. The Windows Azure storage emulator provides local instances of the blob, queue, and table services that are available in Windows Azure. If you are building an application that uses storage services, you can test locally by using the storage emulator.

transient faults. Error conditions that can occur in a distributed environment and that often disappear when you retry the operation. These are often caused by transient problems with the network.

vertical scalability. The ability to increase a computer's resources, such as memory or CPUs.

web role. An interactive application that runs in the cloud. A web role can be implemented with any technology that works with Internet Information Services (IIS) 7.

Windows Azure. Microsoft's platform for cloud-based computing. It is provided as a service over the Internet using either the PaaS or IaaS approaches. It includes a computing environment, the ability to run virtual machines, Windows Azure storage, and management services.

Windows Azure Cloud Services. Web and worker roles in the Windows Azure environment that enable you to adopt the PaaS approach.

Windows Azure Management Portal. A web-based administrative console for creating and managing your Windows Azure hosted services, including Cloud Services, SQL Database, storage, Virtual Machines, Virtual Networks, and Web Sites.

Windows Azure SQL Database. A relational database management system (RDBMS) in the cloud. Windows Azure SQL Database is independent of the storage that is a part of Windows Azure. It is based on SQL Server and can store structured, semi-structured, and unstructured data.

Windows Azure storage. Consists of blobs, tables, and queues. It is accessible with HTTP/HTTPS requests. It is distinct from Windows Azure SQL Database.

Windows Azure Virtual Machine. Virtual machines in the Windows Azure environment that enable you to adopt the IaaS approach.

Windows Azure Virtual Network. Windows Azure service that enables you to create secure site-to-site connectivity, as well as protected private virtual networks in the cloud.

Windows Azure Web Sites. A Windows Azure service that enables you to quickly and easily deploy web sites that use both client and server side scripting, and a database to the cloud.

Worker role. Performs batch processes and background tasks. Worker roles can make outbound calls and open endpoints for incoming calls. Worker roles typically use queues to communicate with Web roles.

Index